MECHANICAL ENGINEERING THEORY AND APPLICATIONS

TOROIDAL SHELLS

Mechanical Engineering Theory and Applications

Additional books in this series can be found on Nova's website under the Series tab.

Additional e-books in this series can be found on Nova's website under the E-book tab.

MECHANICAL ENGINEERING THEORY AND APPLICATIONS

TOROIDAL SHELLS

BOHUA SUN
EDITOR

Nova Science Publishers, Inc.
New York

Library of Congress Cataloging-in-Publication Data

Toroidal shells / editor, Bohua Sun.
p. cm.
Includes index.
ISBN 978-1-61942-247-6 (pbk.)
1. Shells (Engineering)--Mathematical models. 2. Torus (Geometry) I. Sun, Bohua.
TA660.S5T675 2011
624.1'7762--dc23
2011045793

Published by Nova Science Publishers, Inc. † New York

CONTENTS

Preface		**vii**
Chapter 1	Slender Toroidal Shells and Nanotorus *Bohua Sun*	**1**
Chapter 2	Asymptotic Analysis for Toroidal Shells under Arbitrary Loading *R. J. Zhang*	**25**
Chapter 3	Finite Element Analysis for Vibration of Axisymmetric Toroidal Shells of Revolution *George R. Buchanan*	**77**
Chapter 4	A Finite Element Formulation for Piping Structures Based on Thin Shell Displacements Theory *E. M. M. Fonseca, F. J. M. Q. Melo and L. R. Madureira*	**115**
Chapter 5	Vibration of Toroidal Shells and Curved Tubes *Xiaohong Wang*	**151**
Index		**183**

PREFACE

Although research papers on toroidal shells can be found in academic journals, unfortunately, there is no published book dedicated to this subject. This book is the first one in the field, which covers analytical solutions, slender toroidal shells, vibration, pipe system etc. This book is suitable for all engineering under and postgraduates, researchers and scientists.

Chapter 1 - Toroidal shells, in full or partial geometric form, are widely used in structural engineering and have been extensively investigated. Complete toroidal shells spanning 360° in the circumferential and meridian directions are used, for example, as pressure vessels. Such full toroidal shells and curved pipes have received more attention than incomplete toroidal shells spanning less than 360° in the meridian direction. The latter have found use mostly as vehicle or aircraft tyres. Owing to their mechanical advantage, toroidal shells have been proposed for, or used in applications such as underwater missile launchers comprising of large toroidal shells, underwater toroidal pressure hulls, ultra-light and inflated toroidal satellite components, fusion vessels in RFX breeder reactors, rocket fuel tanks and divers' oxygen tanks, circumferential reinforcement for submarines, protective devices for nuclear fuel containers, short and long piping elbows.

Chapter 2 – Toroidal shells are characterized by the existence of the transition points at $\theta = 0$ and $\theta = \pi$. Therefore, displacements at these points are discontinuous in the membrane theory and banding moments always exist. This feature makes it difficult to obtain asymptotic solutions.

Toroidal shell problems are generally divided into two cases to be investigated in accordance with loadings exerted on it: axisymmetric $\left(m=0\right)$ case and non-axisymmetric $\left(m \neq 0\right)$ case, where m is the Fourier harmonic index or the wave number along the circular direction of a toroidal shell.

There are few asymptotic solutions to the case of non-axisymmetric loadings. Steele (1959) solved this problem for the first time in his dissertation at Stanford University. He derived a nonhomogeneous integro-differential equation of fourth order and obtained an asymptotic solution for large values of $\mu = \sqrt{12\left(1-\nu^2\right)} r_0^2 / \left(R_0 h\right)$ and for $m^3 h / \sqrt{12 r_0} \ll 1$. The latter condition restricts this solution adaptable for only lower harmonics: $m = 0, 1, 2, \ldots$.

Chapter 3 - Vibration of shells is a topic that has received the attention of researchers for decades. Early work focused on analysis of the governing equations and exact solutions were dependent upon the ability of the analyst to obtain a solution in the coordinate system describing the shell. Finite difference methods gave researchers another computational tool. But the finite difference method was limited until the advent of computers that enabled an analysis that involved a solution for a significant number of equations. The finite element method gave researchers still another powerful computational tool for the analysis of general shells. The literature that pertains to shell theory is extensive. Toroidal shells have received some attention in the literature, but the coverage is limited when compared with cylindrical and spherical shells. The terminology in this Chapter when referring to thin shells implies that it will mean thin shells with shear deformation and rotary inertia unless clarified to be otherwise.

Chapter 4 - The application of curved pipe elements is critically necessary in any piping structure, no matter which the project applications are. Depending on the safety standards ruling the project, it is known that curved pipe elements are much more vulnerable to excessive stressing than straight pipe elements. A larger risk of accessory breakdown during continuous operation after a situation where limiting forces are exceeded is likely to occur. The study of the flexibility and stress state of curved pipes subjected to generalized forces has been an area of interest of many engineers and physicists, given the high interest of the theme in many structural applications. Practically, in the beginning of 20th Century, important innovative contributions for the research was registered by Theodore von Karman (1911), who proposed the first really effective solution-based on a Fourier expansion displacement field and a variational method.

Chapter 5 - Vibration problems in structures of various shapes and configurations have been extensively studied in the past hundred years or so where many pioneers, Rayleigh and Love in particular, had made considerable contributions in shell structures. Since then, the importance of shell theories can be recognized by the numerous reference books and reviews during the past few decades. In particular, the two distinguished monographs on shell vibration analysis from Leissa (1973) and Soedel (2004) enriched the theories that have broad applications still applicable even today.

In: Toroidal Shells
Editor: Bohua Sun

ISBN: 978-1-61942-247-6

Chapter 1

SLENDER TOROIDAL SHELLS AND NANOTORUS

Bohua Sun
Department of Mechanical Engineering, Cape Peninsula University of Technology, Bellville, Cape Town, South Africa

1. INTRODUCTION

1.1. Slender toroidal shells

Toroidal shells, in full or partial geometric form, are widely used in structural engineering and have been extensively investigated (Reissner 1912; Meissner 1915; Wissler 1916; Zhang 1949; Clark 1950; Clark and Reissner 1951; Clark 1958; Dahl 1953; Novozhilov 1959; Steele 1959; Chien 1979; Zhang et al 1994; Ren et al 1999; Blachut, et al 2000; Redekop and Muhammad 2003). Complete toroidal shells spanning 360° in the circumferential and meridian directions are used, for example, as pressure vessels. Such full toroidal shells and curved pipes have received more attention than incomplete toroidal shells spanning less than 360° in the meridian direction (Redekop and Muhammad 2003). The latter have found use mostly as vehicle or aircraft tyres. Owing to their mechanical advantage, toroidal shells have been proposed for, or used in applications such as underwater missile launchers comprising of large toroidal shells (Ross 2005), underwater toroidal pressure hulls(Blachut 2004), ultra-light and inflated toroidal satellite components (Ruggiero et al 2003), fusion vessels in RFX breeder reactors (Baker et al 2002), rocket fuel tanks and divers' oxygen tanks, circumferential reinforcement for submarines, protective devices for nuclear fuel containers, short and long piping elbows (Wang and Redekop 2005).

There is an extensive literature on the analysis of toroidal shells based on linear elastic theory of thin-walled shells (Sun et al 1990). For different purposes, the formulation of toroidal shells has been obtained by many different shell theories, such as the Sanders-Budiansky shell theory (Sanders Jr 1959, 1963), the Mushtari-Vlasov-Donnell (MVD) (Hoff 1955; Morley 1959; Widera 1986) and the Novozhilov complex transformation shell theory (Novozhilov 1959). Of the available theories, the Mushtari-Vlasov-Donnell (MVD) theory is the simplest to apply. The approximations of the theory have been summarised by

(Novozhilov 1959). The accuracy of the MVD theory, principally for cylindrical shells have been discussed (Hoff 1955; Morley 1959; Widera 1986). Among others, in general, the MDV is considered that the accuracy is greatest for short shells.

Mathematically, the governing equations of toroidal shells are very complicated due to its strong variable confidents and singularity at turning point when $\sin\varphi = -\frac{1}{\alpha} = -\frac{R}{a}$; it is very difficult to find an exact solution and numerical solution valid for whole range of $0 \le \varphi \le 2\pi$. To find their exact solution, the complex form governing equation has been proposed (Reissner 1912; Zhang 1949; Clark et al. 1950, 1958; Dahl 1953; Novozhilov 1959; Steele 1959; Chien 1979; Xia and Zhang 1986; Korovaitsev and Evkin 1992; Kar'yagdyev and Mukoed 1993; Zhang and Zhang 1994; Ren et al 1999), its advantage and disadvantage have been well documented in Novozhilov's master works (Novozhilov 1959). The possibility of turning shells equations into complex form is due to the statical-geometrica analogy, which means that the deformation compatibility equations have complete symmetry with respect to the equilibrium equations of a shell element. It was found that the reduction of the equations to their complex form (without discarding any terms in the equilibrium and compatibility equations) is possible only when using precisely these relations between forces and strains of the middle surface, and then only when Poisson's ratio is equal to zero and the shell thickness is constant. Complex transformation was thus initially extended to any deformation of shells of arbitrary shape only for the particular case when Poisson's ratio is zero. Eventually, after discarding all unimportant terms of the aspect ratio *h/R*, the complex transformation becomes possible for any value of Poisson's ratio μ. The advantage of complex transformation is to reduce the number and order of shells equations. At the same time, complex transformation has some shortcomings: it is not all-encompassing (even within the scope of linear theory) and cannot be extended in general form to include shells of variable thickness and anisotropic shells. It is also not applicable to the dynamic behaviour of shells and to problems of shell stability (buckling). No complex transformation exists in the nonlinear theory of shells (Novozhilov 1962). It is worth noting that the complex form equations of shells is just an approach to reduce the order of the governing equations, unfortunately, there is no universal way on how to derive the complex form equations, every novel complex transformation all claimed to be within the order of thin-shell theory of *h/R*. The reduced complex form equation for toroidal shells can take different forms and has been comprehensively discussed by (Chien 1979).

The most successful application of complex form shell theory is on the study of toroidal shells. The high order and complicated governing equations of toroidal shells under both symmetric and unsymmetrical loads can be reduced to a single equation of low order by complex transformation, and the equation can be solved by the exponential asymptotic method. A very general complex form of toroidal shells equation (Novozhilov 1959) has been formulated as follows:

$$(1+\alpha\sin\varphi)\frac{d^2V}{d\varphi^2} - \alpha\cos\varphi\frac{dV}{d\varphi} + 2\lambda i\sin\varphi V = 2\lambda P_0\cos\varphi$$

$$P_0 = -\frac{1}{2}aq\alpha i + 2\frac{Q_0}{a}\lambda,\ \lambda = \frac{a^2}{2Rh}\sqrt{12\left(1-\mu^2\right)}$$

This equation is a second order ordinary differential equation but with strong variable coefficients, it means that complex transformation can really reduce shell equation's order. Based on the above redcued complex form equation, some successful solutions have been obtained (Novozhilov 1959, Shamina 1962; Zhang and Zhang 1994). Despite its success of finding exact solution for the case of general symmetric toroidal, but for the case of slender toroidal shells ($\alpha = \frac{a}{R} \to 0$), it leads to the following equation:

$d^2V/d^2\varphi + 2i\lambda s \sin\varphi V = 2\lambda P_0 \cos\varphi$ $P_0 = -\frac{1}{2}\alpha\alpha q i$, which still has strong variable coefficient $sin\varphi$ and does not has closed form solution. The closed form solution of the complex-form of governing equation still remains a great challenge for both general and slender toroidal shells.Displacement form governing equations has been used to find a numerical solution instead of an analytical one (Kosawada et al 1985; Zhu 1992; Leung and Kwok 1994; Redekop et al 1999; Ming et al 2002; Fonseca 2002; Jiang and Redekop 2003; Blachut 2003; Wang and Redekop 2004; Redekop 2004; Buchanan et al 2005). There is an extensive studies on various of toroidal shells, such as non-circular cross-section shells (Sutcliffe 1971; Yamada et al 1989; Xu and Redekop 2005; Xu and Redekop 2006; Zhan and Redekop 2007; Redekop 2009), thick toroidal shells (Buchanan and Liu 2005; Redekop 2006) stiffened toroidal shells (Balderes and Armenakas 1973; Wang et al 2006), orthotropic toroidal shells (Bessarabov and Rudis 1966; Xia and Ren 1986; Redekop 2005; Wang and Redekop 2005), plastic loads (Combescure and Galletly 1999; Blachut 2005), plastic instability (Vu and Blachut 2009), buckling and collapse problems (Stein and Mcelman 1965; Weingarten 1973; Wang and Zhang 1991; Galletly and Galletly 1996; Galletly 1998; Blachut and Jaiswal 2000; Blachut 2004; Blachut 2003; Redekop 2005), and post-buckling (Wang and Zhang 1990) and turning point study (Zhang and Zhang 1991).

In the review of all possible literature of toroidal shells, it is very clear that there is no any closed form solution has ever been developed for both general and slender toroidal shells due to their mathematical difficulty. From literature survey, one interesting phenomena is that using complex form equation to solve bending or deformation of toroidal shells, and using displacement to solve vibration and buckling of the shells. For unknown reason, no papers has shown any attempts to deal with the toroidal shells by using displacement form equation, of course, for general toroidal shells, displacement governing is very complicated and very hard to find its exact solution. After carefully study of general displacement form governing equations of toroidal shells, it was surprised to find that the general toroidal equation will be reduced to a simple constant coefficient partial differential equation for slender toroidal shells, and this equation can be easily solved in closed form. In view of the difficulty in the development of solutions for toroidal shells, the closed form solution for slender toroidal shells is indeed an appreciable original contribution to the literature. This work was performed in 1990 when the author was a postdoctoral fellow at Tsinghua University, and then revised recently to meet the latest developments (Sun 1991).

1.2. Nanotours (Figure 10)

Since its discovery, nanotubes have attracted great attention because of their unusual properties [Iijima 1991]. Owing to its exceptional mechanical properties and low density,

Nanotubes are an ideal material for use in nanoeletromechanical systems (NEMS), which are of great interest both for fundamental studies of mechanics at the nanoscale and for a variety of applications[Sun, Bohua and Huang, H XM 2008].

Carbon nanotubes are the strongest and stiffest materials yet discovered in terms of tensile strength and elastic modulus respectively. This strength results from the covalent sp^2 bonds formed between the individual carbon atoms. In 2000, a multi-walled carbon nanotube was tested to have a tensile strength of 63 gigapascals (GPa). This, for illustration, translates into the ability to endure tension of a weight equivalent to 6422 kg on a cable with cross-section of 1 mm^2.) Since carbon nanotubes have a low density for a solid of 1.3 to 1.4 $g{\cdot}cm^{-3}$, its specific strength of up to 48,000 $kN{\cdot}m{\cdot}kg^{-1}$ is the best of known materials, compared to high-carbon steel's 154 $kN{\cdot}m{\cdot}kg^{-1}$. Comparing its strength stress, unfortunately, CNTs are not nearly as strong under compression. Because of their hollow structure and high aspect ratio, they tend to undergo buckling when placed under compressive, torsion or bending stress.

A nanotorus is theoretically described as carbon nanotube bent into a torus (doughnut shape). Nanotori are predicted to have many unique properties, such as magnetic moments 1000 times larger than previously expected for certain specific radii, or may be used as a black body whose emissivity or absorbance is almost of 1.0. Its properties vary widely depending on radius of the torus and radius of the tube. It is expected that it has also has unique mechanical properties.

Comparing the comprehensive studies on the nanotude with cylindrical nanostructures, the nanotorus has not been well investigated[Yakobson 1996; Hod 2003, Huhtala 2002, Cagin 2006, Lourie 1998, Ru 2001]].

Modelling the behaviour of the nanotorus is crucial for device design and interpretation of experimental results. Because experiments at the nanoscale are extremely difficult and atomistic modelling remains prohibitively expensive for large-sized atomic system, it has been accepted that continuum models will continue to play an essential role in the study of nanotorus as a toroidal shells.

The validity of using continuum model for nanotube and graphene has been supported by both experimental results and molecular-dynamics simulation, all previous investigation on nanotube has indicated that the laws of continuum mechanics are still valid to some extent even in nanoscale. The sucessful of continuum model to nanotube gives us a confident to predict that continuum modelling is going to be also valid for nanotori mechanics analysis[67,68.

Due to the development of nanopolymer or nanocomposites, the considerable attention has turned to mechanical behaviour of single walled nanotube embedded in a polymer or metal matrix. The Winkler elastic model will be adopted to study the nanotorus embedded in a elastic medium.

To simplify the toroidal shell modeling, we have noticed a factor that nanotubes have been constructed with length-to-diameter ratio of up to 132,000,000:1, which is significantly larger than any other material. It is means that nanotorus's $\alpha = \frac{a}{R}$ is very small and delectable! Nanotorus can be considered as a very thin toridal shell but not a solid-ring.

Following our previous work in ref.[Sun 2010], after neglecting nanotorus's $\alpha = \frac{a}{R}$, the general toroidal equation will be reduced to a simple constant coefficient partial differential equation for slender toroidal shells, and this equation can be easily solved in closed form. As a natural extension of present works, we also demonstrated a closed-form solution for nanotorus embedded in elastic medium. In view of the difficulty in the development of molecules-dynamics simulation, the simpler solutions obtained here are really useful for the design device.

2. Displacement Type Governing Equations of Toroidal Shells of Symmetric Revolution

From theory of thin shells (Novizhilov, 1959), the rotational symmetric shells of revolution, giving the balancing equations as:

$$\frac{d(rT_1)}{d\varphi} - T_2 R_1 \cos\varphi + N_1 r + q_1 R_1 r = 0$$

$$\frac{d(rN_1)}{d\varphi} - T_1 r - T_2 R_1 \sin\varphi + q_n R_1 r = 0$$

$$\frac{d(rM_1)}{d\varphi} - M_2 R_1 \cos\varphi - N_1 r R_1 = 0 \tag{2.1}$$

The strain displacement relations are given by:

$$\varepsilon_1 = \frac{1}{R_1}(\frac{du}{d\varphi} + w), \ \varepsilon_2 = \frac{1}{r}(u\cos\varphi + w\sin\varphi),$$

$$k_1 = -\frac{1}{R_1}\frac{d}{d\varphi}[\frac{1}{R_1}(\frac{dw}{d\varphi} - u)], \ k_2 = -\frac{\cos\varphi}{rR_1}(\frac{dw}{d\varphi} - u) \tag{2.2}$$

The internal forces and moments are given by:

$$T_1 = K(\varepsilon_1 + \mu\varepsilon_2) = K[\frac{1}{R_1}(\frac{du}{d\varphi} + w) + \frac{\mu}{r}(u\cos\varphi + w\sin\varphi)]$$

$$T_2 = K(\varepsilon_2 + \mu\varepsilon_1) = K[\frac{1}{r}(u\cos\varphi + w\sin\varphi) + \frac{\mu}{R_1}(\frac{du}{d\varphi} + w)]$$

$$M_1 = D(k_1 + \mu k_2) = -D\{\frac{1}{R_1}\frac{d}{d\varphi}[\frac{1}{R_1}(\frac{dw}{d\varphi} - u)] + \frac{\mu\cos\varphi}{rR_1}(\frac{dw}{d\varphi} - u)\}$$

$$M_2 = D(k_2 + \mu k_1) = -D\{\frac{\cos\varphi}{rR_1}(\frac{dw}{d\varphi} - u) + \frac{\mu}{R_1}\frac{d}{d\varphi}[\frac{1}{R_1}(\frac{dw}{d\varphi} - u)]\} \tag{2.3}$$

We can then derive the governing equations for toroidal shells of revolution in terms of displacement components u and w as follows:

$$[r\frac{d}{ds}+(1-\mu)\cos\varphi](\frac{du}{ds}+\frac{w}{R_1})$$
$$+[\mu r\frac{d}{ds}-(1-\mu)\cos\varphi](\frac{u\cos\varphi+w\sin\varphi}{r})$$
$$-\frac{r}{R_1}\frac{D}{K}\frac{d}{ds}(\frac{d}{ds}+\frac{\mu\cos\varphi}{r})(\frac{dw}{ds}-\frac{u}{R_1})$$
$$-\frac{(1-\mu)D}{K}\frac{\cos\varphi}{R_1}(\frac{d}{ds}-\frac{\cos\varphi}{r})(\frac{dw}{ds}-\frac{u}{R_1})+\frac{q_1 r}{K}=0,$$
$$(r\frac{d}{ds}+\cos\varphi)\frac{d}{ds}(\frac{d}{ds}+\frac{\mu\cos\varphi}{r})(\frac{dw}{ds}-\frac{u}{R_1})$$
$$+(1-\mu)(\cos\varphi\frac{d}{ds}-\frac{\sin\varphi}{R_1})(\frac{d}{ds}-\frac{\cos\varphi}{r})(\frac{dw}{ds}-\frac{u}{R_1})$$
$$+\frac{K}{D}[(\frac{r}{R_1}+\mu\sin\varphi)(\frac{du}{ds}+\frac{w}{R_1})$$
$$+(\frac{\mu}{R_1}+\frac{\sin\varphi}{r})(u\cos\varphi+w\sin\varphi)]=\frac{q_n r}{D} \qquad (2.4)$$

where u, w are the displacements, T_1, T_2 are the internal forces, M_1, M_2 are the internal moments, R_1, R_2 are the principal radii of curvature, $ds = R_1 d\varphi$ is the length of the arch, $K=\frac{Eh}{1-\mu^2}$ is the in-plane stiffness, $D=\frac{Eh^3}{12(1-\mu^2)}$ is the bending stiffness, and q_1, q_n are the external loading forces.

3. Displacement Form Governing Equations of Symmetrical Toroidal Shells

The nature of toroidal shells with circular cross-section is that one principal radius is constant and another is variable. From Figures 1, 2 and 3, we have some useful relations as follows:

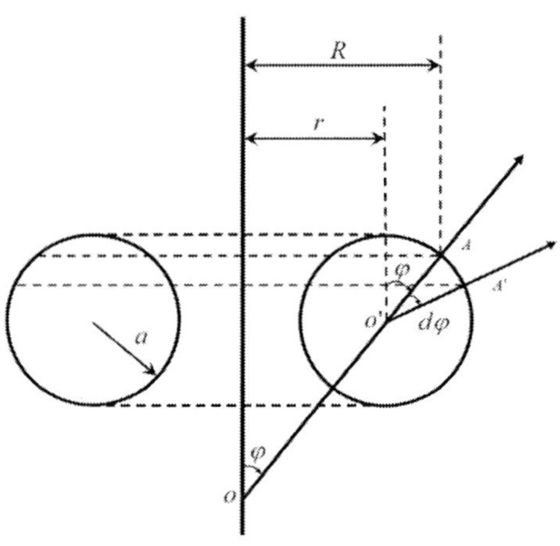

Figure 1. Geometrical description of the toroidal shell.

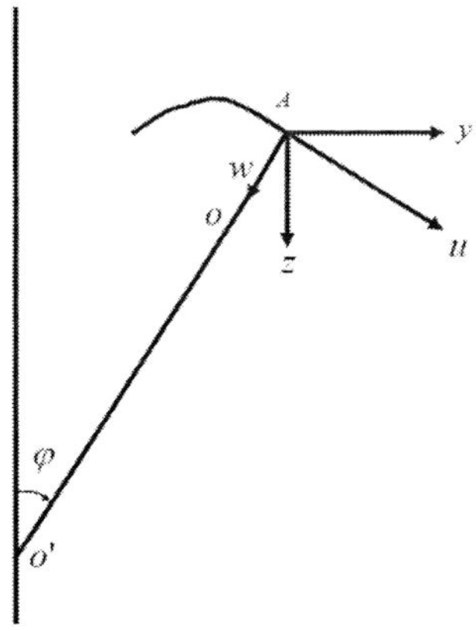

Figure 2. Displacement of the toroidal shell.

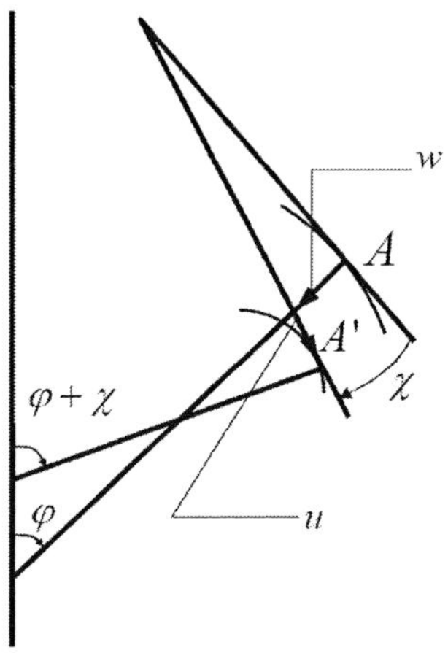

Figure 3. Rotation of normal of toroidal shell.

$$R_1 = a,\ R_2 = \frac{R}{\sin\varphi} + a = \frac{R(1+\alpha\sin\varphi)}{\sin\varphi},$$
$$r = R_2 \sin\varphi = R + a\sin\varphi = R(1+\alpha\sin\varphi),$$
$$\alpha = \frac{a}{R}, \tag{3.1}$$

in which φ is the angle between the symmetric axes and the normal of the toroidal shell, R_1 and R_2 are the radii of curvature of the principal directions of the middle surface, a is the radius of the toroidal shell, and R is the distance from the centre of the toroidal shell to the axes of the shell.

It is important to point out that the second radius $R_2 = \frac{R}{\sin\varphi} + a = \frac{R(1+\alpha\sin\varphi)}{\sin\varphi}$ is function of φ and has two singularity points because it become infinite at points of

$\varphi = \frac{\pi}{2}$ and $\frac{3\pi}{2}$. This singularity is the real source of difficulty of finding exact solution. The novel methods to find exact solution of any toroidal shells are generally taking a strategy in introducing a transformation to remove the singularity.Substituting equation (3.1) into equation (2.3), we have internal forces T_1, T_2 and moments M_1, M_2 of the shells:

$$
\begin{aligned}
T_1 &= K[\frac{1}{a}(\frac{du}{d\varphi}+w)+\mu\frac{u\cos\varphi+w\sin\varphi}{R(1+\alpha\sin\varphi)}],\\
T_2 &= K[\frac{u\cos\varphi+w\sin\varphi}{R(1+\alpha\sin\varphi)}+\frac{\mu}{a}(\frac{du}{d\varphi}+w)],\\
M_1 &= -\frac{D}{a}[\frac{1}{a}\frac{d}{d\varphi}+\frac{\mu\cos\varphi}{R(1+\alpha\sin\varphi)}](\frac{dw}{d\varphi}-u),\\
M_2 &= -\frac{D}{a}[\frac{\cos\varphi}{R(1+\alpha\sin\varphi)}+\frac{\mu}{a}\frac{d}{d\varphi}](\frac{dw}{d\varphi}-u),\\
N_1 &= \frac{1}{a}\frac{dM_1}{d\varphi}+\frac{\cos\varphi}{R}(M_1-M_2),
\end{aligned}
\tag{3.2}
$$

Substituting equation (3.1) into equation (2.4) in appendix, we have the displacement type governing equations for the toroidal shells:

$$
\begin{aligned}
&[\frac{d}{d\varphi}+\frac{(1-\mu)\alpha\cos\varphi}{1+\alpha\sin\varphi}](\frac{du}{d\varphi}+w)\\
&+\alpha[\mu\frac{d}{d\varphi}+\frac{(1-\mu)\alpha\cos\varphi}{1+\alpha\sin\varphi}](\frac{u\cos\varphi+w\sin\varphi}{1+\alpha\sin\varphi})+\frac{q_1a^2}{K}=0,\\
&(\frac{d}{d\varphi}+\frac{\alpha\cos\varphi}{1+\alpha\sin\varphi})\frac{d}{d\varphi}(\frac{d}{d\varphi}+\frac{\alpha\mu\cos\varphi}{1+\alpha\sin\varphi})(\frac{dw}{d\varphi}-u)\\
&+\frac{(1-\mu)\alpha}{1+\alpha\sin\varphi}(\cos\varphi\frac{d}{d\varphi}-\sin\varphi)(\frac{d}{d\varphi}+\frac{\alpha\cos\varphi}{1+\alpha\sin\varphi})(\frac{dw}{d\varphi}-u)\\
&+12(\frac{a}{h})^2[(1+\frac{\alpha\mu\sin\varphi}{1+\alpha\sin\varphi})(\frac{du}{d\varphi}+w)\\
&+\alpha(\mu+\frac{\alpha\sin\varphi}{1+\alpha\sin\varphi})\frac{u\cos\varphi+w\sin\varphi}{1+\alpha\sin\varphi}]=\frac{q_na^4}{D},
\end{aligned}
\tag{3.3}
$$

in which, u and w are the displacement components on the middle surface. Note that we have ignored all terms of $O(h/a)^2$ in deriving the first equation of equations (3.2), since we are using the Love-Kirchhoff theory of shells, in which all terms of order higher than $O(h/a)$ are omitted.

Owing to the complexity of equation (3.3), it is very difficult to find an exact solution for a general ratio of $\alpha = a/R$. In the following section, we will propose a closed form solution of a slender toroidal shell where the ratio is very small and can be omitted. This kind of close form solution has never been reported in the literature.

4. Close Form Solution of Slender Toroidal Axisymmetric Shells

In most of the engineering applications of toroidal shells, there is a common understanding that the ratio $\alpha = \frac{a}{R}$ is much smaller than "1", it means that $\alpha << 1$, and $|\sin\varphi| \le 1$, so that the term $1+\alpha\sin\varphi \approx 1$. It gives a possibility for the simplification and removing its singularity of the toroidal shell equations by ignoring the terms with $\alpha = \frac{a}{R}$ from equation (3). The toroidal shell is called a slender toroidal shell when $\alpha = \frac{a}{R}$ can be omitted. Although we are not able to solve the toroidal shell with arbitrary $\alpha = \frac{a}{R}$, from engineering point of view the slender toroidal shell is a quite good approximation of general toroidal shells. Mathematically, this simplification can be considered as base of perturbation approach if take $\alpha = \frac{a}{R}$ as small perturbation parameter.

Applying $\alpha \to 0$ to equations (3.2) and (3.3), we have equations for the slender toroidal shell as follows:

$$\frac{d}{d\varphi}\left(\frac{du}{d\varphi}+w\right)+\frac{q_1 a^2}{K}=0,$$

$$\frac{d^3}{d\varphi^3}\left(\frac{dw}{d\varphi}-u\right)+12\left(\frac{a}{h}\right)^2\left(\frac{du}{d\varphi}+w\right)=\frac{q_n a^4}{D}, \tag{4.1}$$

These six order ordinary differential equations have a constant coefficient! Its solution will be in closed form. For unknown reasons, the equations have never been derived before. The internal resultant forces and moments are as follows:

$$T_1=\frac{K}{a}\left[\left(\frac{du}{d\varphi}+w\right)+\mu a\underline{\frac{u\cos\varphi+w\sin\varphi}{R}}\right]\approx\frac{K}{a}\left(\frac{du}{d\varphi}+w\right)$$

$$T_2=\frac{K}{a}\left[a\underline{\frac{(u\cos\varphi+w\sin\varphi)}{R}}+\mu\left(\frac{du}{d\varphi}+w\right)\right]\approx\frac{\mu K}{a}\left(\frac{du}{d\varphi}+w\right)$$

$$M_1=-\frac{D}{a^2}\left[\frac{d}{d\varphi}+\underline{\frac{\mu a\cos\varphi}{R}}\right]\left(\frac{dw}{d\varphi}-u\right)\approx-\frac{D}{a^2}\frac{d}{d\varphi}\left(\frac{dw}{d\varphi}-u\right)$$

$$M_2=-\frac{D}{a^2}\left[\mu\frac{d}{d\varphi}+\underline{\frac{a\cos\theta}{R}}\right]\left(\frac{dw}{d\varphi}-u\right)\approx-\frac{\mu D}{a^2}\frac{d}{d\varphi}\left(\frac{dw}{d\varphi}-u\right)$$

$$N_1=\frac{1}{a}\left[\frac{dM_1}{d\varphi}+\underline{\frac{a\cos\varphi}{R}(M_1-M_2)}\right]\approx-\frac{D}{a^3}\frac{d^2}{d\varphi^2}\left(\frac{dw}{d\varphi}-u\right) \tag{4.2}$$

Equation (4.1) can be further simplified by letting:

$$\psi = \frac{du}{d\varphi} + w, \qquad \chi = \frac{dw}{d\varphi} - u \tag{4.3}$$

where χ is the rotation of normal vector of the middle surface. Then, from the 1st equation of equation (4.1), we have:

$$\psi(\varphi) = \frac{a^2}{K}[C_1 - \int_\theta q_1(\xi)d\xi] \tag{4.4}$$

where C_1 is an integration constant.

Substituting equation (4.4) into the second equation of (4.1), we have the rotation as follows:

$$\chi = C_2 + C_3\varphi + C_4\varphi^2 + \frac{a^4}{D}\int_\theta (\varphi-\xi)^2 q_n(\xi)d\xi - 12(\frac{a}{h})^2 \int_\theta (\varphi-\xi)^2 \psi(\xi)d\xi \tag{4.5}$$

Using equation (4.4), then equation (4.5) can be rewritten as follows:

$$\chi(\varphi) = -2C_1(\frac{a}{h})^2 \frac{a^2}{K}\varphi^3 + C_2 + C_3\varphi + C_4\varphi^2 + \frac{a^4}{D}\int_\theta (\varphi-\xi)^2 q_n(\xi)d\xi + 12(\frac{a}{h})^2 \frac{a^2}{K}\int_\theta (\varphi-\xi)^3 q_1(\xi)d\xi \tag{4.6}$$

where C_2, C_3, C_4 are integration constants.

From equation (4.3), we have:

$$\frac{d^2u}{d\varphi^2} + u = \frac{d\psi}{d\varphi} - \chi \tag{4.7}$$

$$\frac{d^2w}{d\varphi^2} + w = \frac{d\chi}{d\varphi} + \psi \tag{4.8}$$

As long as we find a solution to any one of equation (4.7) and (4.8), we will solve the problem. The closed form solution of equation (4.7) is as follows:

$$u = C_5\cos\varphi + C_6\sin\varphi + \int_\varphi p(\xi)\sin(\varphi-\xi)d\xi \tag{4.9}$$

where

$$p(\varphi)=\frac{d\psi}{d\varphi}-\chi$$
$$=-\frac{q_1 a^2}{K}+2C_1(\frac{a}{h})^2\frac{a^2}{K}\varphi^3-C_2-C_3\varphi-C_4\varphi^2$$
$$-\frac{a^4}{D}\int_\varphi(\varphi-\xi)^2 q_n(\xi)d\xi-12(\frac{a}{h})^2\frac{a^2}{K}\int_\varphi(\varphi-\xi)^3 q_1(\xi)d\xi \qquad (4.10)$$

From equation (4.3), we have deflection as follows:

$$w=\psi-\frac{du}{d\varphi}=\frac{a^2}{K}[C_1-\int_\varphi q_1(\xi)d\xi]+C_5\sin\varphi$$
$$-C_6\cos\varphi-\int_\varphi p(\xi)\cos(\varphi-\xi)d\xi \qquad (4.11)$$

Substituting (4.9) and (4.11) into equation (4.2), we can get internal forces and moments in terms of displacement u and w. The six integration constants C_i (i = 1,...,6) can be determined by using boundary conditions.

5. Toroidal Shells Used as Thermal Expansion Compensation Devices

Figure 4 shows a typical thermal expansion compensation device, in which the toroidal shell is used as the expansion mechanism. The mechanism is loaded by an axial force P. We will apply our solution to the problem to find the expansion of the mechanism under this loading condition.

This problem has been studied by using complex form equation (Dahl 1953; Qian 1979). Unfortunately, they only use its particular solution to study the problem and never derived its homogenous solution. Both of them did not provide a comprehensive study of the problem. We present here a complete rational study as an application our slender toroidal shells. In this problem, since $q_1=q_n=0$, we have displacement components as follows:

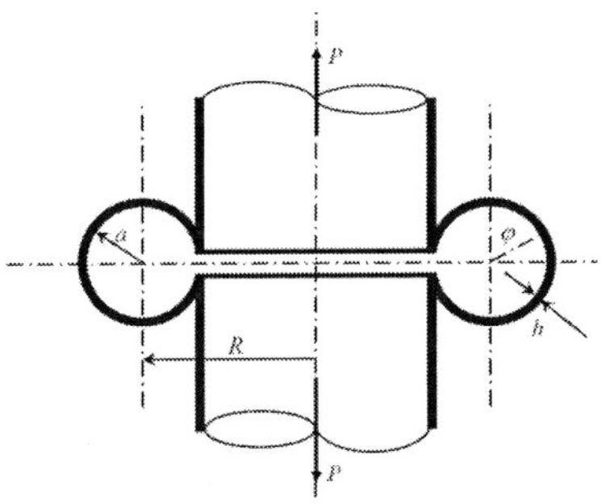

Figure 4. Toroidal shell as a thermal expansion compensation device.

$$u = C_5 \cos\varphi + C_6 \sin\varphi + \int_\varphi p(\xi)\sin(\varphi - \xi)d\xi$$

$$w = \frac{a^2}{K}C_1 + C_5 \sin\varphi - C_6 \cos\varphi - \int_\varphi p(\xi)\cos(\varphi - \xi)d\xi \tag{5.1}$$

where $p(\varphi) = \frac{d\psi}{d\varphi} - \chi = 2C_1(\frac{a}{h})^2 \frac{a^2}{K}\varphi^3 - C_2 - C_3\varphi - C_4\varphi^2$. After completion of integration, then we have the displacements of the thermal devices as follows:

$$u = C_5 \cos\varphi + C_6 \sin\varphi + 2C_1(\frac{a}{h})^2 \frac{a^2}{K}\varphi^3 - C_4\varphi^2 - C_3\varphi - C_2$$

$$w = \frac{a^2}{K}C_1 + C_5 \sin\varphi - C_6 \cos\varphi - 2C_1(\frac{a}{h})^2 \frac{a^2}{K}(3\varphi^3 + 6) + C_3 + 2C_4\varphi \tag{5.2}$$

and its rotations:

$$\psi(\varphi) = \frac{a^2}{K}C_1 \text{ and } \chi = -2C_1(\frac{a}{h})^2 \frac{a^2}{K}\varphi^3 + C_2 + C_3\varphi + C_4\varphi^2 \tag{5.3}$$

The internal forces and moments are given by:

$$T_1 = \frac{K}{a}(\frac{du}{d\varphi} + w) = \frac{K}{a}\psi = aC_1$$

$$M_1 \approx -\frac{D}{a^2}\frac{d\chi}{d\varphi} = -\frac{D}{a^2}[-6C_1(\frac{a}{h})^2 \frac{a^2}{K}\varphi^2 + C_3 + 2C_4\varphi]$$

$$N_1 \approx -\frac{D}{a^3}\frac{d^2\chi}{d\varphi^2} = -\frac{D}{a^3}[-12C_1(\frac{a}{h})^2 \frac{a^2}{K}\varphi + 2C_4] \tag{5.4}$$

For the toroidal shell, the following boundary conditions apply:

$$\varphi = \frac{\pi}{2}, \quad u = 0, \ \text{w=0}, \ \chi\text{=0}$$

$$\varphi = -\frac{\pi}{2}, \quad \text{T}_1 = -\frac{P}{2\pi(R-a)} \approx -\frac{P}{2\pi R}, \quad \text{N}_1\text{=0}, \ \text{M}_1\text{=0} \tag{5.5}$$

We then have the following integration constants:

$$C_1 = -\frac{P}{2\pi aR}, \ C_2 = -\frac{7\pi^2}{8}(\frac{a}{h})^2 \frac{a}{R}\frac{P}{K}, \ C_3 = \frac{3\pi}{4}(\frac{a}{h})^2 \frac{a}{R}\frac{P}{K}, \ C_4 = \frac{3}{2}(\frac{a}{h})^2 \frac{a}{R}\frac{P}{K}$$

$$C_5 = -(\frac{6}{\pi} + \frac{9\pi}{4} + \frac{3\pi^2}{8})(\frac{a}{h})^2 \frac{a}{R}\frac{P}{K} + \frac{1}{2\pi}\frac{a}{R}\frac{P}{K}, C_6 = 0.,$$

Substituting the integration constants then gives final internal forces and moments:

$$T_1 = \frac{K}{a}(\frac{du}{d\varphi} + w) = \frac{K}{a}\psi = -\frac{P}{2\pi R}$$

$$M_1 \approx -\frac{D}{a^2}\frac{d\chi}{d\varphi} = -\frac{3}{\pi}\frac{D}{a^2}(\frac{a}{h})^2\frac{a}{R}\frac{P}{K}(\varphi + \frac{\pi}{2})^2$$

$$N_1 \approx -\frac{D}{a^3}\frac{d^2\chi}{d\varphi^2} = -\frac{6}{\pi}\frac{D}{a^3}(\frac{a}{h})^2\frac{a}{R}\frac{P}{K}(\varphi + \frac{\pi}{2}) \tag{5.6}$$

and the displacement field:

$$u = [-(\frac{6}{\pi} + \frac{9\pi}{4} + \frac{3\pi^2}{8})(\frac{a}{h})^2\frac{a}{R}\frac{P}{K} + \frac{1}{2\pi}\frac{a}{R}\frac{P}{K}]\cos\varphi + (\frac{a}{h})^2\frac{a}{R}\frac{P}{K}(\frac{7\pi^2}{8} - \frac{3\pi}{4}\varphi - \frac{3}{2}\varphi^2 - \frac{1}{\pi}\varphi^3)$$

$$w = [-(\frac{6}{\pi} + \frac{9\pi}{4} + \frac{3\pi^2}{8})(\frac{a}{h})^2\frac{a}{R}\frac{P}{K} + \frac{1}{2\pi}\frac{a}{R}\frac{P}{K}]\sin\varphi - \frac{1}{2\pi}\frac{a}{R}\frac{P}{K} + (\frac{a}{h})^2\frac{a}{R}\frac{P}{K}(\frac{6}{\pi} + \frac{3\pi}{4} + 3\varphi + \frac{3}{\pi}\varphi^3) \tag{5.7}$$

and the rotation:

$$\chi = (\frac{a}{h})^2\frac{a}{R}\frac{P}{K}(\frac{1}{\pi}\varphi^3 + \frac{3}{2}\varphi^2 + \frac{3\pi}{4}\varphi - \frac{7\pi^2}{8}) \tag{5.8}$$

We then have the expansion displacement under the action of the axial loading P as follows:

$$\delta = 2u_{\varphi=-\frac{\pi}{2}} = 2\pi^2(\frac{a}{h})^2\frac{a}{R}\frac{P}{K} = 2\pi^2\frac{(1-\nu^2)}{R}\frac{Pa^3}{Eh^3} \tag{5.9}$$

If using bending stiffness given by $D = Eh^3/[12(1-\mu^2)]$, the expansion displacement can be rewritten as:

$$\delta_{Sun} = 2u_{\varphi=-\frac{\pi}{2}} = 2\pi^2\frac{12(1-\mu^2)}{Eh^3}\frac{Pa^3}{12R} = \frac{\pi^2}{6}\frac{Pa^3}{RD} = 1.644934\frac{Pa^3}{RD} \tag{5.10}$$

This solution indicates that the weaker the bending stiffness, the greater the expansion displacement.

In comparing the above solution with the solution obtained (Dahl 1953; Qian 1979), they used several approximations to simplify the equation and proposed a solution as follows (which can also be found in Roark's Formulas for Stress and Strain 2002):

$$\delta_{Dahl} = 2u_{\varphi=-\frac{\pi}{2}} = \begin{cases} \sqrt{12(1-\mu^2)}\dfrac{Pa}{Eh^2} & \text{if } \sqrt{3(1\text{-}\mu^2)}\,\frac{a^2}{Rh} \to \infty \\ \dfrac{1}{2}\dfrac{Pa^3}{RD} & \text{if } \sqrt{3(1\text{-}\mu^2)}\,\frac{a^2}{Rh} \to 0 \end{cases} \tag{5.11}$$

It is clear that our solution is valid for all cases, and does not need to specify the range of the parameter as in equation (5.11).

6. Nanotorus Embedded in Elastic Medium (Winkler Model)

When the nanotube or nanotorus embedded in composite materials, the nanotorus-matric interaction can be viewed as a nanotorus buried in a elastic foundation, which can be described as a Winkler model. There are some investigations on nanotude buried in the elastic medium[7,8], unfortunately, there is no one paper has considered the case of nanotorus embedded in elastic medium. As a natural extension of this works, we will present closed form solution of the nanotorus embedded in Winkler elastic medium.

In Winkler model, the loading distribution in normal direction of the nanotorus will be in the form of $q_n = q - kw$, where k is the Winkler constant, which is determined by the material properties of the elastic matrix and curvature of the nanotorus. Here, the dependency of the constant k on the curvature can be neglected as a second-order effect.

After introducing the Winkler model, the governing equation of the nanotorus embedded in elastic medium take following form:

$$\begin{aligned} &\frac{d}{d\varphi}\left(\frac{du}{d\varphi} + w\right) = 0, \\ &\frac{d^3}{d\varphi^3}\left(\frac{dw}{d\varphi} - u\right) + 12\left(\frac{a}{h}\right)^2\left(\frac{du}{d\varphi} + w\right) + k\frac{a^4}{D}w = \frac{qa^4}{D}, \end{aligned} \tag{6.1}$$

In which, team $\dfrac{q_1 a^2}{K}$ has be removed in the first equation of (6.1) since we are not consider the friction between nanotorus and matrix. The we have

$$w = C_1 - \frac{du}{d\varphi}, \tag{6.2}$$

$$u = \sum_{i=3}^{6} C_i e^{\lambda_i \varphi} + \frac{D}{ka^4}\left[-\frac{qa^4}{D}\varphi + 12\left(\frac{a}{h}\right)^2 C_1\varphi + k\frac{a^4}{D}C_1\varphi + C_2\right], \tag{6.3}$$

where $\lambda_i = \pm\sqrt{\dfrac{-1 \pm \sqrt{1 - 4\dfrac{ka^4}{D}}}{2}}$.

7. Vibration of Nanotorus

The advantage of using the displacement type equations is that we are able to study the deformation as well as the vibration of the shells. For literal vibration, the vibration in the direction of u can be ignored due to its relatively small value. Then equation becomes, by deleting the inertial term $\frac{\rho h a^2}{K}\ddot{u}$:

$$\frac{\partial}{\partial\varphi}\left(\frac{\partial u}{\partial\varphi}+w\right)=0$$

$$\frac{\partial^3}{\partial\varphi^3}\left(\frac{\partial w}{\partial\varphi}-u\right)+12\left(\frac{a}{h}\right)^2\left(\frac{\partial u}{\partial\varphi}+w\right)+\frac{\rho h a^4}{D}\ddot{w}=0 \tag{7.1}$$

where ρ is the density of the shell material, and the partial derivative with respect to time $\dot{u}=\frac{\partial u}{\partial t}$. The solution is given by:

$$u(\varphi,t)=[b_3e^{\beta_1\varphi}+b_4e^{\beta_2\varphi}+b_5e^{\beta_3\varphi}+b_6e^{\beta_4\varphi}-\frac{1}{\omega^2}\frac{D}{\rho h a^4}(\frac{12a^2}{h^2}b_1\varphi+b_2)](B_1\cos\omega t+B_2\sin\omega t) \tag{7.2}$$

$$\begin{aligned} w=b_1-[&b_3\beta e_1^{\beta_1\varphi}+b_4\beta_2e^{\beta_2\varphi}+b_5\beta_3e^{\beta_3\varphi}+b_6\beta_4e^{\beta_4\varphi}\\ &-\frac{12}{\omega^2}\frac{D}{\rho h^3a^2}b_1](B_1\cos\omega t+B_2\sin\omega t)\end{aligned} \tag{7.3}$$

where $b_i(i=1,2,3,4,5,6)$ are integration constants, and

$$\beta_i=\pm\sqrt{\frac{-1\pm\sqrt{1+4\frac{\rho h a^4}{D}\omega^2}}{2}}.$$

With the solution of the free vibration problem, that for forced vibration can be derived by using the superposition principal of vibration modes. This is a standard operation which is not difficult.

8. Buckling of Nanotorus

General speaking, the buckling of toroidal shells is a very difficult problem, and no analytical solution has been obtained. By taking advantage of the displacement form of the governing equations for toroidal shells, we can find its analytical solution of the buckling problem.

The buckling equation of the slender toroidal shell as follows:

$$\frac{d}{d\varphi}(\frac{du}{d\varphi}+w)=0$$

$$\frac{d^3}{d\varphi^3}(\frac{dw}{d\varphi}-u)+12(\frac{a}{h})^2(\frac{du}{d\varphi}+w)$$

$$+\frac{T_1^*}{a}\frac{d}{d\varphi}(\frac{dw}{d\varphi}-u)+T_2^*\frac{\cos\varphi}{R}(\frac{dw}{d\varphi}-u)=0 \tag{8.1}$$

where T^*_1, T^*_2 are the internal forces for the zero-moment state of the shells, which satisfies the membrane balance equation $\frac{T_1^*}{R_1}+\frac{T_2^*}{R_2}=q_n$. The first equation of equations (8.1) becomes $\frac{du}{d\varphi}+w=-\frac{1}{12}(\frac{a}{h})^2 d_1$, d_1 is an integration constant. The second equation of equations (8.1) becomes:

$$\frac{d^3\chi}{d\varphi^3}+\frac{T_1^*}{a}\frac{d\chi}{d\varphi}+T_2^*\frac{\cos\varphi}{R}\chi=d_1 \tag{8.2}$$

This is the buckling equation for slender toroidal shells, where rotation of normal vector of the nanotorus $\chi=\frac{dw}{d\varphi}-u$.

It is worth noting that the strain $\varepsilon_1=\frac{1}{a}(\frac{du}{d\varphi}+w)=-\frac{1}{12}\frac{a}{h^2}d_1$ is a constant and is independent of q_n, this is a special feature of nanotorus. For a nanotorus undergoing a distributed compressive loading, $T_1^*=aq_n, T_2^*=\frac{1}{2}aq_n$, equation (8.2) becomes to:

$$\frac{d^3\chi}{d\varphi^3}+q_n\frac{d\chi}{d\varphi}+\underline{\tfrac{1}{2}(\frac{a}{R}\cos\varphi)\chi}=d_1 \tag{8.3}$$

The underlined term can be ignored due to the fact that the ratio $\alpha=\frac{a}{R}$ is very small. Then the buckling equation for a slender toroidal shell becomes:

$$\frac{d}{d\varphi}(\frac{d^2\chi}{d\varphi^2}+q_n\chi)=d_1 \quad \text{or} \quad \frac{d^2\chi}{d\varphi^2}+q_n\chi=d_1\varphi+d_2 \tag{8.4}$$

Its solution can be easily found by the theory of ordinary differential equations:

$$\chi = d_3 \cos\sqrt{q_n}\varphi + d_4 \sin\sqrt{q_n}\varphi + \frac{1}{q_n}(d_1\varphi + d_2) \tag{8.5}$$

where d_i are integration constants. And the displacements are as follows:

$$\begin{aligned} u &= d_5 \cos\varphi + d_6 \sin\varphi \\ &\quad -\frac{1}{1-q_n} d_3 \cos\sqrt{q_n}\varphi - \frac{1}{1+q_n} d_4 \sin\sqrt{q_n}\varphi - \frac{1}{q_n}(d_1\varphi + d_2) \\ w &= [\frac{1}{q_n} - \frac{1}{12}(\frac{a}{h})^2]d_1 + d_5 \sin\varphi - d_6 \cos\varphi \\ &\quad - d_3 \frac{\sqrt{q_n}}{1-q_n} \sin\sqrt{q_n}\varphi + d_4 \frac{\sqrt{q_n}}{1+q_n} \cos\sqrt{q_n}\varphi \end{aligned} \tag{8.6}$$

The internal forces and moments are given by:

$$\begin{aligned} T_1 &= \frac{K}{a}(\frac{du}{d\varphi} + w) = -\frac{1}{12}\frac{a}{h^2} d_1 \\ M_1 &= -\frac{D}{a^2}\frac{d\chi}{d\varphi} = -\frac{\sqrt{q_n}D}{a^2}[-d_3\sqrt{q_n} \sin\sqrt{q_n}\varphi + d_4\sqrt{q_n} \cos\sqrt{q_n}\varphi + \frac{1}{q_n} d_1] \\ N_1 &= -\frac{D}{a^2}\frac{d^2\chi}{d\varphi^2} = \frac{q_n D}{a^2}[d_3 \cos\sqrt{q_n}\varphi + d_4 \sin\sqrt{q_n}\varphi] \end{aligned} \tag{8.7}$$

Using proper boundary conditions, the critical value of q_n can be obtained as an eigenvalue of the problem.

9. Numerical Demonstration of the Thermal Device

As a numerical application of the above closed form solution for the thermal devices, we will demonstrate three case of $\frac{a}{R}$ ratio. All data are taken from paper (Wang and Redekop 2005) and shown in Table 1. All results are demonstrated in the respective figure.

Table 1. Parameters for numerical calculation (data taken from Wang and Redekop 2005)

	Unit	Case 1	Case 2	Case 3
a	[m]	0.05	0.1	0.2
R	[m]	1.0	1.0	1.0
H	[m]	0.004	0.004	0.004
E	Pa [N/m^2]	0.207 x 10^{12}	0.207 x 10^{12}	0.207 x 10^{12}
μ		0.3	0.3	0.3
P	[Kg]	1000	1000	1000

Figure 5 shows the moment M_1 (Unit is Kg.m)in the toroidal shell reaches its maximum at $\varphi = \frac{\pi}{2}$, and will decrease with the increasing change in *a*. Figure 6 shows the resultant force N_1 (Unit is Kg) in the toroidal shell reaches its maximum at $\varphi = \frac{\pi}{2}$, and will not change with the change of *a,* it means the resultant forces are in same value for the three case. It has a linear relationship with φ. Figure 7 shows the displacement *u* (Unit is m) in the toroidal shell reaches its maximum at $\varphi = -\frac{\pi}{2}$ and has a zero point between φ= -1 and 0. It is interesting to note that the zero point of the displacement is independent of the change of toroidal radius *a*. Figure 8 shows the deflection *w* (Unit is m) in the toroidal shell reaches its maximum at about $\varphi = -1$ and has a zero point between φ= 1 and 0. As in the case of the displacement, zero point of the deflection is independent of the change of toroidal radius *a*. Figure 9 shows that the slope of toroidal shell is "zero" at $\varphi = \pi/2$ and maximum at $\varphi = -\pi/2$.

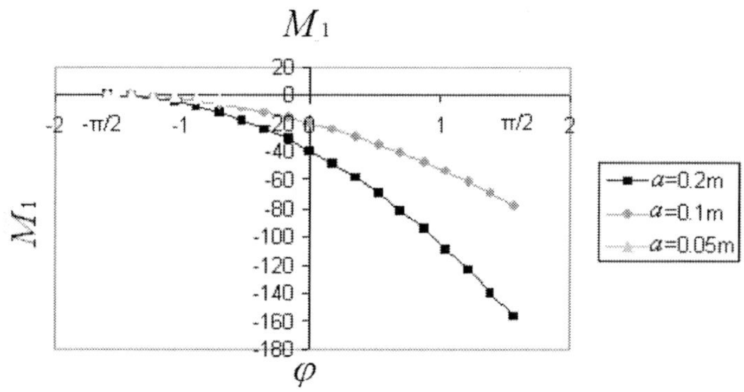

Figure 5. The moment M_1(Unit is Kgm) of the shell.

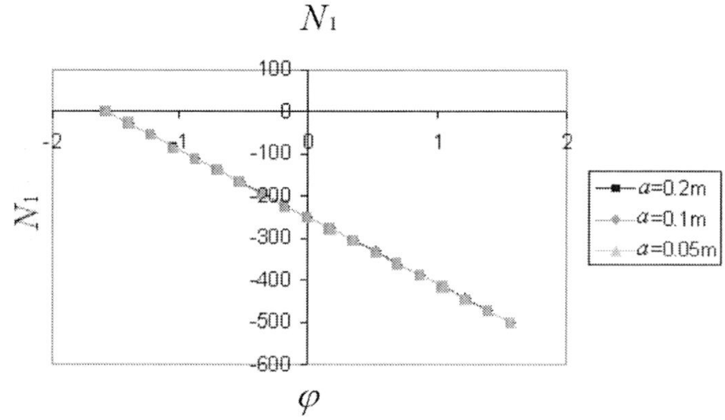

Figure 6. The resultant force N_1 (Unit is Kg) of middle surface of the shell.

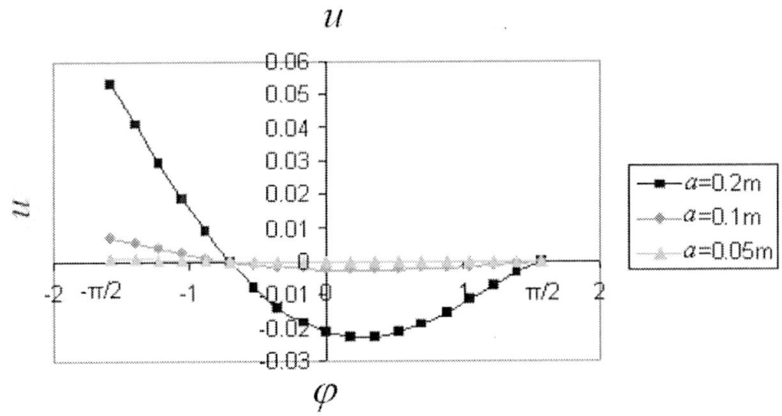

Figure 7. The displacement u (Unit is m) in middle surface of the shell.

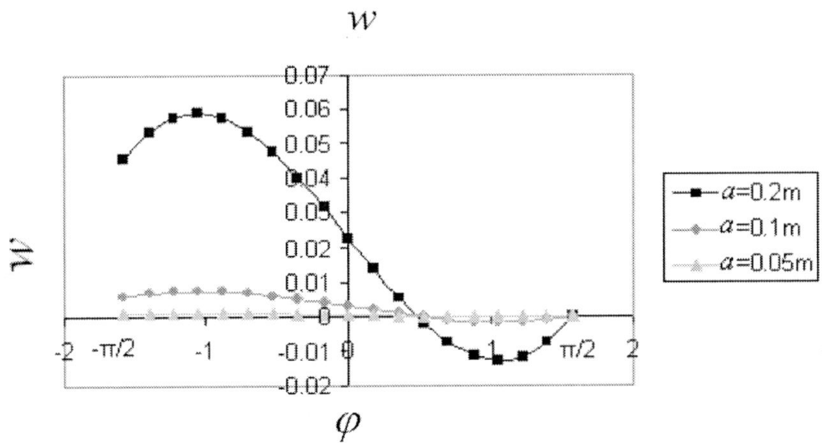

Figure 8. The defection w (Unit is m) of the shell.

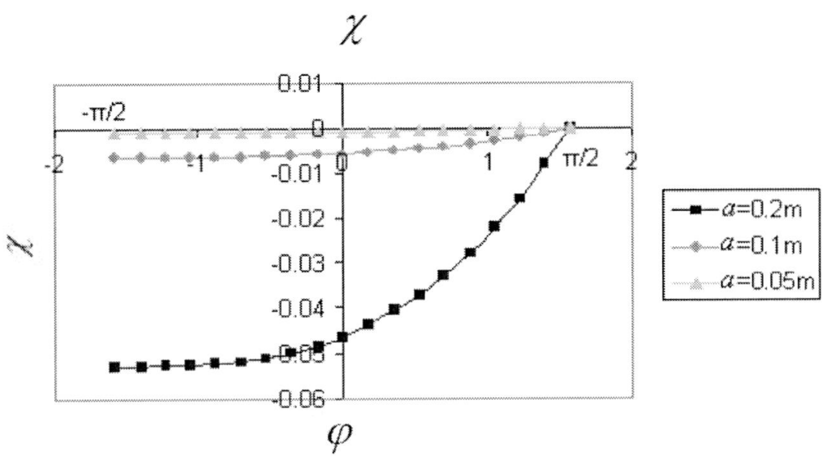

Figure 9. The slope χ of the shell.

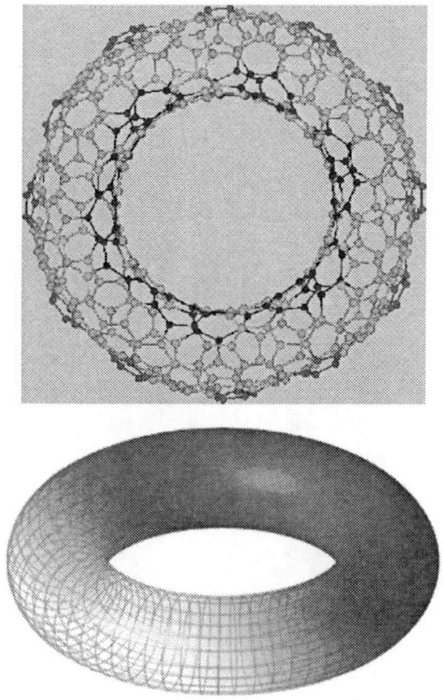

Figure 10. Complete nanotours and its continuum model.

CONCLUSION

1) The slender toroidal shell solution is a very good approximation of general toroidal shells in the case of $\alpha = \frac{a}{R} \ll 1$.
2) The slender toroidal shell is a good macro mechanical continuum model of nanotorus.
3) For slender toroidal shell, the displacement form equation is much simpler than complex form equation.
4) The solution presented by (Dahl 1953) for thermal expansion devices has to be modified and replaced by our solution (25), given as

$$\delta_{Sun} == \frac{\pi^2}{6}\frac{Pa^3}{RD} = 1.644934\frac{Pa^3}{RD}.$$

5) The displacement equations can be easily extended to the problems of vibration and buckling of the shells, those will be studied in the forthcoming papers.

ABOUT AUTHOR

Bohua Sun has been with Cape Peninsula University of Technology (CPUT) since 1995, where he is currently Professor and director of the Centre for Mechanics, Smart Structures and Microsystems. He is a member of both Academy of Science of South Africa (ASSAf) and Royal Society of South Africa.

REFERENCES

Baker, W.R., Bello, S.D., Marcuzzi, D., Sonato, P., Zaccaria, P. (2002). "Design of a new toroidal shell and support for RFX." *Fusion Eng Des,* 64:461–6.

Balderes, T., Armenakas, A.E. (1973). "Free vibrations of ring-stiffened toroidal shells." *AIAA Journal,* 11:1637–1644.

Bessarabov, Yu D. and Rudis, M.A. (1966). "On the symmetrical deformation of an orthotropic toroidal shell." *Proceedings of the 4th All-Union Conference on Shells and Plates*, Erevan 24-31 October 1962, pp.207-215, edited by S.M. Dugar'yan (English translation), *Israel Programm for Scientific Translations*, Jerusalem, (1966).

Blachut, J. (2003). "Collapse tests on externally pressurised toroids", *J. Pressure Vessel Technology, Transactions of the ASME,* Vol., 125, 91-96.

Blachut, J. (2004). "Buckling and first ply failure of a composite toroidal pressure hull." *Comput Struct*, 82:1981–92.

Blachut, J. (2005). "Plastic loads for internally pressurised toroidal shells." *J. Pressure Vessel Technology, Transactions of the ASME,* Vol., 127, 2005, 151-156.

Blachut, J., Jaiswal O.R. (2000). "On the buckling of toroirdal shells under external pressure." *Comp Struct,* 77:233–51.

Bohua Sun, Closed-Form Solution of Axisymmetric Slender Elastic Toroidal Shells, *J. Engrg. Mech*. Volume 136, Issue 10, pp. 1281-1288 (October 2010)

Buchanan, G.R., Liu Y.J. (2005). "An analysis of the free vibration of thick-walled isotropic toroidal shells." *International Journal of Mechanical Science,* 47:277–292.

ÇAĞIN1, T., et al, Computational studies on mechanical properties of carbon nanotori, *Turkish Journal of Physics*, 30, 221-229 (2006).

Chien, W.Z. (1979). "Selected Papers in Applied Math and Mechanics." *Jiangsu Sciences and Technology Press.*

Clark, R. A. (1958). "Asymptotic solutions of toroidal shell problems." *Quart. Appl.. Math.*, 16, 47-60.

Clark, R. A. and Reissner, E. (1951). "Bending of curved tubes." *Advances in Applied Mechanics*, Vol.,2. 93-122.

Clark, R.A. (1950). "On the theory of thin elastic toroidal shells." *J. Mech. Phys.*, Vol.,29, No.3, 146-178.

Combescure, A., Galletly G.D. (1999). "Plastic buckling of complete toroidal shells of elliptical cross-section subjected to internal pressure." *Thin-Walled Struct.* 34:135–46.

Dahl, N.C. (1953). "Toroidal-shell expansion joints." *ASME J. of Applied Mechanics*, 20, 497-503.

Galletly, G.D. (1998). "Elastic buckling of complete toroidal shells of elliptical cross-section subjected to uniform internal pressure." *Thin-Walled Struct.,* 30:23–34.

Galletly, G.D., Galletly D.A. (1996). "Buckling of complex toroidal shell structures." *Thin Wall Struct*, 26:195–212.

Hod, O. and Rabani, E., Cabon nanotube closed-ring structures. *Physical Review, B* 67, 195408 (2003).

Hoff, N. J. (1955). "The accuracy of Donnell's equations." *J. Appl. Mech.,* 22, 329-334.

Huhtala, M. *et al*, Carbon nanotube structures: molecular dynamics simulation at realistic limit. *Computer Physics Communications,* 146, 30-37 (2002).

Iijima, S., Helical microtubules of graphitic carbon. *Nature,* 354, 56 - 58 (07 November 1991).

Jiang, W., Redekop D. (2003). "Static and vibration analysis of orthotropic toroidal shells of variable thickness by differential quadrature." *Thin-Walled Structures*, 41:461–478.

Kar'yagdyev, N. Ya., and Mukoed A. P. (1993). " Axisymmetric deformation of flexible orthotropic toroidal shells." *Journal of Mathematical Sciences*, Vol., 66, No. 5.

Korovaitsev, A.V., Evkin, A.Iu. (1992). "Axisymmetrical deformation of a toroidal shell under strong bending." *Prikladnaia Mekhanika* (ISSN 0032-8243), Vol., 28, No. 4, p. 16-23. (in Russian).

Leung A.Y.T., Kwok N.T.C., (1994). "Free vibration analysis of a toroidal shell." *Thin Wall Struct*, 18:317–32.

Lourie, O., *et al.* Buckling and collapse of embedded carbon nanotubes. *Phys. Rev. Lett.* 81, 1638–1641 (1998).

Meissner, E. (1915). "Uber und Elastizitat Festigkeit dunner Schalen." *Viertelschr. D. nature.Ges*., Bd.60, Zurich.

Ming, R.S., Pan J., Norton M.P. (2002). "Free vibrations of elastic circular toroidal shells." *Applied Acoustics*, 63:513–528.

Morley, L.S.D. (1959). "An improvement on Donnell's approximation for thin-walled circular cylinders." *Quart. J. Mech. Appl. Math*., 12, 89-99.

Novozhilov, V.V. (1959). "The Theory of Thin Shells." *Noordhoff, Groningen.*

Novozhilov, V.V. (1966). "Development of the method of complex transformation in the linear theory of shells during the last fifty years." *Proceedings of the 4th All-Union Conference on Shells* and Plates held at Erevan 24-31 October 1962, pp.88-95, edited by S.M. Dugar'yan (English translation), *Israel Programm for Scientific Translations*, Jerusalem.

Pomares, R.J., Durlofsky, H. (1988). "Collapse analysis of toroidal shell." *ASME PVP,* 199:23–33.

Redekop, D. (2004). "Free vibration of hollow bodies of revolution." *Journal of Sound and Vibration*, 273: 415–420.

Redekop, D. (2005). "Buckling analysis of an orthotropic thin shell of revolution using differential quadrature." *Int. J. Press Ves. Piping.,* 82:618–24.

Redekop, D. (2006). "Three-dimensional free vibration analysis of inhomogenous thick orthotropic shells of revolution using differential quadrature." *Journal of Sound and Vibration*, 291: 1029-1040.

Redekop, D. (2009). "Buckling analysis of an orthotropic elliptical toroidal shell" *Proceedings of the ASME 2009 Pressure Vessels and Piping Division Conference*, PVP2009, July 26-30, 2009, Prague, Czech Republic, PVP2009-77524, ASME, pp. 1-8, 2009 (ISBN 978-0-7918-3854-9).

Redekop, D., Muhammad, T. (2003). "Analysis of toroidal shells using the differential quadrature method." *Int. J. of Structural Stability and Dynamics,* 3:215–226.

Reissner, H. (1912). Spannungen in Kugelschalen (Kuppeln), Müller-Breslau Festschrift, 181–193 Leipzig.

Ren, W., Liu, W., Zhang, W., Reimerdes, H.G. and Oery, H. (1999). "A survey of works on the theory of toroidal shells and curved tubes." *Acta Mechanica Sinica*, Vol., 15, No. 3, 225-234.

Ross, C.T.F. (2005). "A conceptual design of an underwater missile launcher." *Ocean Eng* 32:85–99.

Ru, CQ Axially compressed buckling of a double walled nanotube embedded within an elastic medium. *Journal of the Mechanics and Physics of Solids,* 49 1265-1279, (2001).

Ruggiero, E.J., Jha, A., Park, G., Inman, D.J. (2003). "A literature review of ultra-light and inflated toroidal satellite components." *Shock Vib Digest*, 35:171–81.

Sanders, Jr J.L. (1959). "An improved first-approximation theory for thin shells." *NASA Tech. Rep.*, R-24:1–11.

Sanders, Jr. JL. (1963). "Nonlinear theories for thin shells." *Q Appl Math.,* 21, 21–36.

Shamina, V.A. (1962). "Calculation of a ribbed toroidal shell under a symmetrical road." *Proceedings of the 4th All-Union Conference on Shells and Plates* held at Erevan 24-31 October 1962, pp.907-911, edited by S.M. Dugar'yan (English translation), *Israel Programm for Scientific Translations, Jerusalem,* (1969).

Steele, C. R. (1959). "Toroidal shells with nonsymmetric loading." *Ph.D. Dissertation*, Stanford University.

Stein, M, Mcelman, J.A. (1965). "Buckling of Segments of toroidal shells." *AIAA J.*, 3:1704-1709, 1965.

Sun, B. (1991). "Progress in Applied Mechanics." *Postdoctoral Research Report*, Tsinghua University, Beijing, China.

Sun, Bohua and Huang, H XM, Mechanical nano-resonators at ultra-high frequency and their potential applications, *South Africa Journal of Science*, vol., 104, pages 169-171, (2008).

Sutcliffe, W.J. (1971). "Stress analysis of toroidal shells of elliptical cross section." *Int. J. Mech. Sci.*, 13:951–8.

Vu, V.T., Blachut, J. (2009). "Plastic instability pressure of toroidal shells." *J. Pressure Vessel Technology, Transactions of the ASME*, Vol. 131(5), 0512031-05120310.

Wang, A., Zhang, W. (1990). "Solution for postbuckling of toroidal shells." *Sci. China, Ser.* A, 33(10):1220-1229.

Wang, A., Zhang, W. (1991). "Asymptotic solution for buckling of toroidal shells." *Int. J. Pres. Ves. Piping*, 45:61-72.

Wang, X.H., Redekop, D.(2005). "Natural frequencies and mode shapes of an orthotropic thin shell of revolution." *Thin-Walled Structures,* 43:733–750.

Wang, X.H., Xu, B., Redekop, D. (2006). "FEM free vibration and buckling analysis of stiffened toroidal shells." *Thin-Walled Structures*, 44:2–9.

Wang, X.H., Xu, B., Redekop, D. (2006). "Theoretical natural frequencies and mode shapes for thin and thick curved pipes and toroidal shells." *Journal of Sound and Vibration*, 292:424-434.

Weingarten, V.I., Veronda, D.R., Saghera, S.S. (1973). "Buckling of segments of toroidal shells." *AIAA J.*, 11:1422-1424.

Widera, G.E.O. (1986). "Validity of various shell theories applicable in the design and analysis of cylindrical pressure vessels." *ASME PVP*, 100: 11-19.

Wissler, H. (1916). "Festigkeiberechung von Ringsflachen." *Promotionarbeit*, Zurich.

Xia, Z. H. and Zhang W. (1986). “The general solution for thin-walled curved tubes with arbitrary loadings and various boundary conditions.” *Int. J. Pressures and Piping,* 26, 129-144.

Xia, Z., and Ren, W. (1986). “An analysis of stress and strain for orthotropic toroidal shells.” *Nuclear Engineering and Design/Fusion*, 3: 309-318.

Xu, B., Redekop, D. (2006). “Natural frequencies of an orthotropic thin toroidal shell of elliptical cross-section.” *J. of Sound Vibration*, 293:440–448.

Yakobson, B. I., Brabec, C. J., and Bernholc, J., Nanomechanics of carbon tubes: Instabilities beyond linear response. Phys. Rev. Lett., 76, 2511(1996).

Yamada, G., Kobayahsi, Y., Ohta, Y., Yokota S. (1989). “Free vibration of a toroidal shell with elliptical cross-section.” *J. of Sound Vibration*, 135:411–25.

Young, W.C., Budynas, R.G.. (2002). “Roark's Formulas for Stress and Strain (7th Edition).” McGraw-Hill.

Zhan, H.J., Redekop, D. (2008). “Vibration, buckling and collapse of ovaloid toroidal tank.” *Thin-Walled Structures*, 46: 380-389.

Zhang, R. J. and Zhang, W. (1991). “Turning point solution for thin shell vibrations.” *Int. J. Solids Structures*, 27(10), 1311-1326.

Zhang, R.J. and Zhang, W. (1994). “Toroidal Shells under Nonsymmetric Loading.” *Int. J. Solids Structures*, Vol. 31, No. 19. 2735-2750.

In: Toroidal Shells
Editor: Bohua Sun

ISBN: 978-1-61942-247-6

Chapter 2

ASYMPTOTIC ANALYSIS FOR TOROIDAL SHELLS UNDER ARBITRARY LOADING

R. J. Zhang*
School of Aerospace Engineering and Applied Mechanics,
Tongji University, Shanghai, China

NOMENCLATURE

The following symbols are used in this paper:

h = wall thickness of shells;

r_0 = radius of meridian circle;

R_0 = distance from top of toroid to axis of revolution;

$\mu = r_0 / R_0$;

m = Fourier harmonic index;

$$\lambda^2 = \begin{cases} i\sqrt{12\left(1-\nu^2\right)}\dfrac{\mu r_0}{h} m^4 & (\text{for non-axisymmetric}) \\ i\sqrt{12\left(1-\nu^2\right)}\dfrac{\mu r_0}{h} & (\text{for axisymmetric}) \\ i\sqrt{12\left(1-\nu_1\nu_2\right)\dfrac{E_2}{E_1}\dfrac{\mu r_0}{h}} & (\text{axisymmetric for orthotropic}) \end{cases}$$: a complex large parameter;

z = Langer's variable;

$\zeta = -\lambda^{2/3} z$: fast varying variable;

$\psi^{(k)}$ = k th homogeneous solutions ($k=4$ for non-axisymmetric, $k=2$ for axisymmetric);

$\psi_0^{(k)}$ = 0 order approximation to homogeneous solutions;

$\psi_{high}^{(k)}$ = high order approximation to homogeneous solutions;

---

* e-mail:zhangrj@tongji.edu.cn.

ψ^* = particular solution;

ψ_0^* = 0 order approximation to particular solution;

ψ_{high}^* = high order approximation to particular solution;

$\phi_1^{(0)}, \phi_2^{(0)}$ = two homogeneous solutions of membrane equation in non-axisymmetric case;

$\phi_*^{(0)}$ = particular solutions of membrane equation in non-axisymmetric case;

$A_k(\zeta, p), B_k(\zeta, p), B_0(\zeta, p)$ = General Airy functions.

1. Introduction

Toroidal shells are characterized by the existence of the transition points at $\theta = 0$ and $\theta = \pi$. Therefore, displacements at these points are discontinuous in the membrane theory and banding moments always exist. This feature makes it difficult to obtain asymptotic solutions.

Toroidal shell problems are generally divided into two cases to be investigated in accordance with loadings exerted on it: axisymmetric $(m = 0)$ case and non-axisymmetric $(m \neq 0)$ case, where m is the Fourier harmonic index or the wave number along the circular direction of a toroidal shell.

There are few asymptotic solutions to the case of non-axisymmetric loadings. Steele (1959) solved this problem for the first time in his dissertation at Stanford University. He derived a nonhomogeneous integro-differential equation of fourth order and obtained an asymptotic solution for large values of $\mu = \sqrt{12(1-\nu^2)}\, r_0^2/(R_0 h)$ and for $\mathrm{m^3 h}/\sqrt{12\mathrm{r_0}} \ll 1$. The latter condition restricts this solution adaptable for only lower harmonics: $m = 0, 1, 2, \ldots$.

The axisymmetric problem of toroidal shells has been studied by many investigators based on different equations. W.Zhang (1944, 1949), R.A.Clark (1950), V.V. Novozhilov (1951) and C.A.Tumarkin (1959) gave a first approximation to the homogeneous solutions, respectively, in terms of or Bessel's functions or Airy functions.

Tumarkin (1959) obtained a complete asymptotic expansion for the particular solution in terms of Lommel functions, which arrives at the accuracy of the thin shell theory. Years later, his result was justified by Clark (1963).

In the present paper, we start from a set of equations in complex form based on Sanders' thin shell theory.

It is well known that Novozhilov was the first to formulate such equations for a shell with an arbitrary middle surface subjected to arbitrarily distributed loads.

From the point of view of accuracy, there is little difference among all the existing sets of shell equations. However, there are two features in Sanders' shell theory: 1) the equations can be written in general tensor form for arbitrary shells. 2) The theory contains an exact static-geometric analogy. In fact, for toroidal shells, Sanders' equations are identical in form to Novozhilov's equations. This is why we may use all the equations given by Novozhilov for the axisymmetric case.

In the present paper, all the asymptotic solutions for arbitrarily distributed loads, including axisymmetric and non-axisymmetric cases, are given in terms of general Airy functions. These asymptotic solutions are uniformly valid in whole toroidal shell including transition points. For all the solutions, not only the first approximation but high approximation is given. Therefore, all the solutions arrive at the accuracy of thin shell theory. It is worthy to indicates that all the solutions, including homogeneous and particular solutions in both axisymmetric and non-axisymmetric cases, are expressed in terms of general Airy functions instead of or Airy functions or Bessel's functions for homogeneous solutions and Lommel functions for the particular solutions as in the other papers.

General Airy functions are introduced by Drazin and Reid (1981) for fluid stability problem. Using the general Airy functions the author gave a set of asymptotic solutions for the transition point problem in free vibrations of shells of revolution in 1986.

2. Toroidal Shell Equations in Complex Form

In Sanders' theory for thin shells the eight force and moment resultants N_{11}, $\tilde{N}_{12}$, N_{22}, Q_1, Q_2, M_{11}, $\tilde{M}_{12}$ and M_{22} satisfy the following equilibrium equations:

$$\frac{\partial \alpha_2 N_{11}}{\partial \xi_1}+\frac{\partial \alpha_1 \tilde{N}_{12}}{\partial \xi_2}+\frac{\partial \alpha_1}{\partial \xi_2}\tilde{N}_{12}-\frac{\partial \alpha_2}{\partial \xi_1}N_{22}+\frac{\alpha_1\alpha_2}{R_1}Q_1+\frac{\alpha_2}{2}\frac{\partial}{\partial \xi_2}\left[\left(\frac{1}{R_1}-\frac{1}{R_2}\right)\tilde{M}_{12}\right]=0 \tag{2-1}$$

$$\frac{\partial \alpha_2 \tilde{N}_{12}}{\partial \xi_1}+\frac{\partial \alpha_1 N_{22}}{\partial \xi_2}+\frac{\partial \alpha_2}{\partial \xi_1}\tilde{N}_{12}-\frac{\partial \alpha_1}{\partial \xi_2}N_{11}+\frac{\alpha_1\alpha_2}{R_2}Q_2+\frac{\alpha_2}{2}\frac{\partial}{\partial \xi_1}\left[\left(\frac{1}{R_2}-\frac{1}{R_1}\right)\tilde{M}_{12}\right]=0 \tag{2-2}$$

$$\frac{\partial \alpha_2 Q_1}{\partial \xi_1}+\frac{\partial \alpha_1 Q_2}{\partial \xi_2}-\alpha_1\alpha_2\left(\frac{N_{11}}{R_1}+\frac{N_{22}}{R_2}\right)=0 \tag{2-3}$$

$$\frac{\partial \alpha_2 M_{11}}{\partial \xi_1}+\frac{\partial \alpha_1 \tilde{M}_{12}}{\partial \xi_2}+\frac{\partial \alpha_1}{\partial \xi_2}\tilde{M}_{12}-\frac{\partial \alpha_2}{\partial \xi_1}M_{22}-\alpha_1\alpha_2 Q_1=0 \tag{2-4}$$

$$\frac{\partial \alpha_2 \tilde{M}_{12}}{\partial \xi_1}+\frac{\partial \alpha_1 M_{22}}{\partial \xi_2}+\frac{\partial \alpha_2}{\partial \xi_1}\tilde{M}_{12}-\frac{\partial \alpha_1}{\partial \xi_2}M_{11}-\alpha_1\alpha_2 Q_2=0 \tag{2-5}$$

where $\tilde{N}_{12}=\frac{1}{2}\left(N_{12}+N_{21}\right)$ (2-6)

•

$$\text{and } \tilde{M}_{12} = \frac{1}{2}\left(M_{12} + M_{21}\right) \tag{2-7}$$

are the modified force resultant and the modified moment resultant, respectively. All the 10 force and moment resultants in the above equations, N_{11}, N_{12}, N_{21}, N_{22}, Q_1.

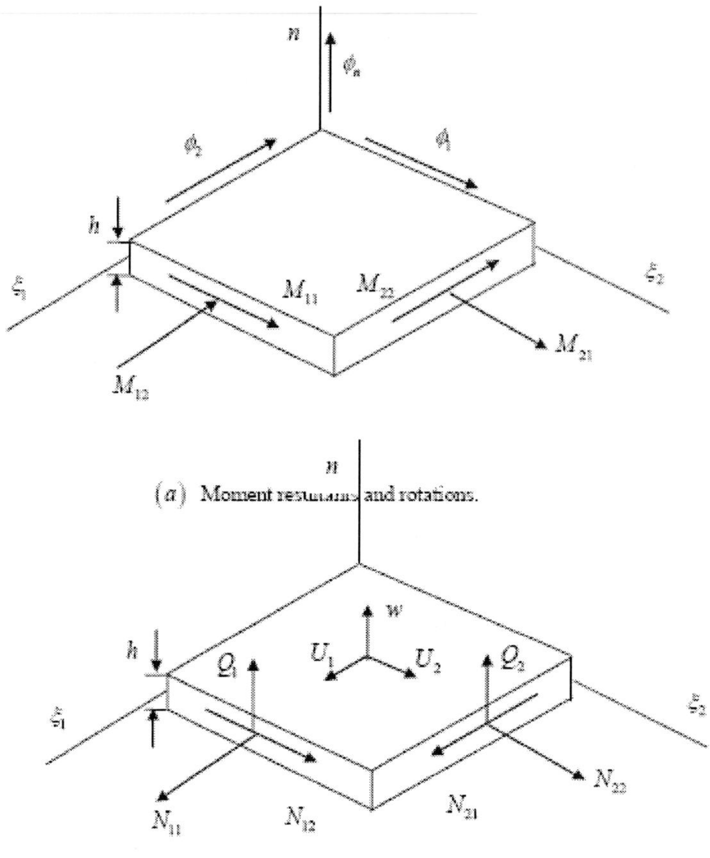

Figure 2.1. Orientation of coordinates, displacements, rotations, force resultants and moment resultants.

Q_2, M_{11}, M_{12}, M_{21} and M_{22}, which act on sections of the shell parallel to the coordinate curves ξ_1 and ξ_2, are shown in Figure 2.1.

The three strains $\varepsilon_{11}, \varepsilon_{12}, \varepsilon_{22}$ and three curvatures $\kappa_{11}, \kappa_{22}, \tilde{\kappa}_{12}$ satisfy the following compatibility equations:

$$\frac{\partial \alpha_2 \kappa_{22}}{\partial \xi_1} - \frac{\partial \alpha_1 \tilde{\kappa}_{12}}{\partial \xi_2} - \frac{\partial \alpha_1}{\partial \xi_2}\tilde{\kappa}_{12} - \frac{\partial \alpha_2}{\partial \xi_1}\kappa_{11} + \frac{\alpha_1 \alpha_2}{R_1}P_1 + \frac{\alpha_2}{2}\frac{\partial}{\partial \xi_2}\left[\left(\frac{1}{R_1} - \frac{1}{R_2}\right)\varepsilon_{12}\right] = 0 \tag{2-8}$$

$$\frac{\partial \alpha_1 \kappa_{11}}{\partial \xi_2} - \frac{\partial \alpha_2 \tilde{\kappa}_{12}}{\partial \xi_1} - \frac{\partial \alpha_2}{\partial \xi_1}\tilde{\kappa}_{12} - \frac{\partial \alpha_1}{\partial \xi_2}\kappa_{22} + \frac{\alpha_1 \alpha_2}{R_2}P_2 + \frac{\alpha_2}{2}\frac{\partial}{\partial \xi_1}\left[\left(\frac{1}{R_2} - \frac{1}{R_1}\right)\varepsilon_{12}\right] = 0 \tag{2-9}$$

$$\frac{\partial \alpha_2 P_1}{\partial \xi_1} + \frac{\partial \alpha_1 P_2}{\partial \xi_2} - \alpha_1 \alpha_2 \left(\frac{\kappa_{11}}{R_1} + \frac{\kappa_{22}}{R_2} \right) = 0 \tag{2-10}$$

$$-\frac{\partial \alpha_2 \varepsilon_{22}}{\partial \xi_1} + \frac{\partial \alpha_1 \varepsilon_{12}}{\partial \xi_2} + \frac{\partial \alpha_1}{\partial \xi_2} \varepsilon_{12} + \frac{\partial \alpha_2}{\partial \xi_1} \varepsilon_{11} - \alpha_1 \alpha_2 P_1 = 0 \tag{2-11}$$

$$-\frac{\partial \alpha_1 \varepsilon_{11}}{\partial \xi_2} + \frac{\partial \alpha_2 \varepsilon_{12}}{\partial \xi_1} + \frac{\partial \alpha_2}{\partial \xi_1} \varepsilon_{12} + \frac{\partial \alpha_1}{\partial \xi_2} \varepsilon_{22} - \alpha_1 \alpha_2 P_2 = 0 \tag{2-12}$$

in which P_1 and P_2 are two auxiliary functions introduced by the author. In the compatibility equations the six strain quantities are defined in terms of displacements and rotations as

$$\varepsilon_{11} = \frac{1}{\alpha_1} \frac{\partial U_1}{\partial \xi_1} + \frac{1}{\alpha_1 \alpha_2} \frac{\partial \alpha_1}{\partial \xi_2} U_2 + \frac{W}{R_1} \tag{2-13}$$

$$\varepsilon_{22} = \frac{1}{\alpha_2} \frac{\partial U_2}{\partial \xi_2} + \frac{1}{\alpha_1 \alpha_2} \frac{\partial \alpha_2}{\partial \xi_1} U_1 + \frac{W}{R_2} \tag{2-14}$$

$$\varepsilon_{12} = \frac{1}{2\alpha_1 \alpha_2} \left(\alpha_2 \frac{\partial U_2}{\partial \xi_1} + \alpha_1 \frac{\partial U_1}{\partial \xi_2} - \frac{\partial \alpha_1}{\partial \xi_2} U_1 - \frac{\partial \alpha_2}{\partial \xi_1} U_2 \right) \tag{2-15}$$

$$\kappa_{11} = \frac{1}{\alpha_1} \frac{\partial \phi_1}{\partial \xi_1} + \frac{1}{\alpha_1 \alpha_2} \frac{\partial \alpha_1}{\partial \xi_2} \phi_2 \tag{2-16}$$

$$\kappa_{22} = \frac{1}{\alpha_2} \frac{\partial \phi_2}{\partial \xi_2} + \frac{1}{\alpha_1 \alpha_2} \frac{\partial \alpha_2}{\partial \xi_1} \phi_1 \tag{2-17}$$

$$\tilde{\kappa}_{12} = \frac{1}{2\alpha_1 \alpha_2} \left[\alpha_2 \frac{\partial \phi_2}{\partial \xi_1} + \alpha_1 \frac{\partial \phi_1}{\partial \xi_2} - \frac{\partial \alpha_1}{\partial \xi_2} \phi_1 - \frac{\partial \alpha_2}{\partial \xi_1} \phi_2 + \frac{1}{2} \left(\frac{1}{R_2} - \frac{1}{R_1} \right) \left(\frac{\partial \alpha_2 U_2}{\partial \xi_1} - \frac{\partial \alpha_1 U_1}{\partial \xi_2} \right) \right] \tag{2-18}$$

where U_1, U_2 and W are displacements ; ϕ_1 and ϕ_2 are the rotations of the shell, which can be expressed in terms of displacements as

$$\phi_1 = \frac{U_1}{R_1} - \frac{1}{\alpha_1} \frac{\partial W}{\partial \xi_1} \tag{2-19}$$

$$\phi_2 = \frac{U_2}{R_2} - \frac{1}{\alpha_2} \frac{\partial W}{\partial \xi_2} \tag{2-20}$$

All the displacements and rotations are shown in Figure 2.1.

In the case of an isotropic elastic material, the stress-strain relations are taken to be the form:

$$Eh\varepsilon_{11} = N_{11} - \nu N_{22} \quad \frac{Eh^3}{12}\kappa_{11} = M_{11} - \nu M_{22}$$

$$Eh\varepsilon_{22} = N_{22} - \nu N_{11} \quad \frac{Eh^3}{12}\kappa_{22} = M_{22} - \nu M_{11} \tag{2-21}$$

$$Eh\varepsilon_{12} = (1+\nu)\tilde{N}_{12} \quad \frac{Eh^3}{12}\bar{\kappa}_{12} = (1+\nu)\tilde{M}_{12}$$

where h is the shell thickness, E is Young's modulus, and ν is Poisson's ratio.

Sanders indicated in 1959 that if in the compatibility equations (2-8) to (2-15) ε_{11} is replaced by $-M_{22}$, ε_{22} is replaced by $-M_{11}$, ε_{12} is replaced by $\bar{M}_{12}$, κ_{11} is replaced by N_{22}, κ_{22} is replaced by N_{11}, $\tilde{\kappa}_{12}$ is replaced by $-\tilde{N}_{12}$, P_1 is replaced by Q_1, and P_2 is replaced by Q_2, then these equations become identical to the equations (2-1) to (2-5).

Thus, a set of complex variables can then be introduced as

$$N_1 = N_{11} - iEhC\kappa_{22}, \ N_2 = N_{22} - iEhC\kappa_{11}, \ S = \tilde{N}_{12} + iEhC\tilde{\kappa}_{12} \tag{2-22$_1$}$$

$$M_1 = M_{11} + iEhC\varepsilon_{22}, \ M_2 = M_{22} + iEhC\varepsilon_{11},$$

$$H = \tilde{M}_{12} - iEhC\varepsilon_{12} \tag{2-22$_2$}$$

$$F_1 = Q_1 + iEhCP_1, \ F_2 = Q_2 + iEhCP_2 \tag{2-22$_3$}$$

where $i = \sqrt{-1}$ is the imaginary unit and

$$\varepsilon = \frac{h}{\sqrt{12(1-\nu^2)}} \tag{2-23}$$

Multiplying the compatibility equations (2-8) to (2-15) in terms of $-iEh\varepsilon$, and adding them to the equilibrium equations (2-1) to (2-5), respectively, give

$$\frac{\partial \alpha_2 N_1}{\partial \xi_1} + \frac{\partial \alpha_1 S}{\partial \xi_2} + \frac{\partial \alpha_1}{\partial \xi_2} S - \frac{\partial \alpha_2}{\partial \xi_1} N_2 + \frac{\alpha_1 \alpha_2}{R_1} F_1 + \frac{\alpha_2}{2}\frac{\partial}{\partial \xi_2}\left[\left(\frac{1}{R_1} - \frac{1}{R_2}\right) H\right] = 0 \tag{2-24}$$

$$\frac{\partial \alpha_2 S}{\partial \xi_1}+\frac{\partial \alpha_1 N_2}{\partial \xi_2}+\frac{\partial \alpha_2}{\partial \xi_1}S-\frac{\partial \alpha_1}{\partial \xi_2}N_1+\frac{\alpha_1\alpha_2}{R_2}F_2+\frac{\alpha_2}{2}\frac{\partial}{\partial \xi_1}\left[\left(\frac{1}{R_2}-\frac{1}{R_1}\right)H\right]=0 \tag{2-25}$$

$$\frac{\partial \alpha_2 F_1}{\partial \xi_1}+\frac{\partial \alpha_1 F_2}{\partial \xi_2}-\alpha_1\alpha_2\left(\frac{N_1}{R_1}+\frac{N_2}{R_2}\right)=0 \tag{2-26}$$

$$\frac{\partial \alpha_2 M_1}{\partial \xi_1}+\frac{\partial \alpha_1 H}{\partial \xi_2}+\frac{\partial \alpha_1}{\partial \xi_2}H-\frac{\partial \alpha_2}{\partial \xi_1}M_2-\alpha_1\alpha_2 F_1=0 \tag{2-27}$$

$$\frac{\partial \alpha_2 H}{\partial \xi_1}+\frac{\partial \alpha_1 M_2}{\partial \xi_2}+\frac{\partial \alpha_2}{\partial \xi_1}H-\frac{\partial \alpha_1}{\partial \xi_2}M_1-\alpha_1\alpha_2 F_2=0 \tag{2-28}$$

Combining (2-21) with (2-22) gives

$$M_1=i\varepsilon\left(N_2-\nu\bar{N}_1\right),\ M_2=i\varepsilon\left(N_1-\nu\bar{N}_2\right),\ H=-i\varepsilon\left(S+\nu\bar{S}\right) \tag{2-29}$$

Hereafter a "bar operation" always denotes conjugate. So that $\bar{N}_1$, $\bar{N}_2$ and $\bar{S}$ are the conjugate counterpart to N_1, N_2 and S, respectively. Then M_1, M_2 and H in (2-24) and (2-25) can be eliminated by use of a substitution from (2-29). The result is

$$\frac{\partial \alpha_2 N_1}{\partial \xi_1}+\frac{\partial \alpha_1 S}{\partial \xi_2}+\frac{\partial \alpha_1}{\partial \xi_2}S-\frac{\partial \alpha_2}{\partial \xi_1}N_2+\frac{\alpha_1\alpha_2}{R_1}F_1-i\varepsilon\frac{\alpha_2}{2}\frac{\partial}{\partial \xi_2}\left[\left(\frac{1}{R_1}-\frac{1}{R_2}\right)\left(S+\nu\bar{S}\right)\right]=0$$

$$\frac{\partial \alpha_2 S}{\partial \xi_1}+\frac{\partial \alpha_1 N_2}{\partial \xi_2}+\frac{\partial \alpha_2}{\partial \xi_1}S-\frac{\partial \alpha_1}{\partial \xi_2}N_1+\frac{\alpha_1\alpha_2}{R_2}F_2-i\varepsilon\frac{\alpha_2}{2}\frac{\partial}{\partial \xi_1}\left[\left(\frac{1}{R_2}-\frac{1}{R_1}\right)\left(S+\nu\bar{S}\right)\right]=0$$

It is easily found that the last terms in the two equations are small quantities of the order of $\frac{\varepsilon}{R}\sim\frac{h}{R}$ (R is the smallest radius of curvature of shells) in comparison with the terms containing S in the same equation. Neglecting them yields

$$\frac{\partial \alpha_2 N_1}{\partial \xi_1}+\frac{\partial \alpha_1 S}{\partial \xi_2}+\frac{\partial \alpha_1}{\partial \xi_2}S-\frac{\partial \alpha_2}{\partial \xi_1}N_2+\frac{\alpha_1\alpha_2}{R_1}F_1=0 \tag{2-30}$$

$$\frac{\partial \alpha_2 S}{\partial \xi_1}+\frac{\partial \alpha_1 N_2}{\partial \xi_2}+\frac{\partial \alpha_2}{\partial \xi_1}S-\frac{\partial \alpha_1}{\partial \xi_2}N_1+\frac{\alpha_1\alpha_2}{R_2}F_2=0 \tag{2-31}$$

Additionally, substituting (2-29) into (2-27) and (2-28) yields

$$\frac{\alpha_1\alpha_2}{R_1}F_1 = i\frac{\varepsilon}{R_1}\left[\frac{\partial\alpha_2 N_2}{\partial\xi_1} - \frac{\partial\alpha_1 S}{\partial\xi_2} - \frac{\partial\alpha_1}{\partial\xi_2}S - \frac{\partial\alpha_2}{\partial\xi_1}N_1 - \nu\left(\frac{\partial\alpha_2\bar{N}_1}{\partial\xi_1} + \frac{\partial\alpha_1\bar{S}}{\partial\xi_2} + \frac{\partial\alpha_1}{\partial\xi_2}\bar{S} - \frac{\partial\alpha_2}{\partial\xi_1}\bar{N}_2\right)\right] \tag{2-32}$$

$$\frac{\alpha_1\alpha_2}{R_2}F_2 = -i\frac{\varepsilon}{R_2}\left[\frac{\partial\alpha_2 S}{\partial\xi_1} - \frac{\partial\alpha_1 N_1}{\partial\xi_2} + \frac{\partial\alpha_2}{\partial\xi_1}S + \frac{\partial\alpha_1}{\partial\xi_2}N_2 + \nu\left(\frac{\partial\alpha_2\bar{S}}{\partial\xi_1} + \frac{\partial\alpha_1\bar{N}_2}{\partial\xi_2} + \frac{\partial\alpha_2}{\partial\xi_1}\bar{S} - \frac{\partial\alpha_1}{\partial\xi_2}\bar{N}_1\right)\right] \tag{2-33}$$

Two notations which permit the above equations to be written in a more compact form are introduced as follows:

$$L_1 = \frac{\partial\alpha_2 N_1}{\partial\xi_1} + \frac{\partial\alpha_1 S}{\partial\xi_2} + \frac{\partial\alpha_1}{\partial\xi_2}S - \frac{\partial\alpha_2}{\partial\xi_1}N_2 \tag{2-34}$$

$$L_2 = \frac{\partial\alpha_2 S}{\partial\xi_1} + \frac{\partial\alpha_1 N_2}{\partial\xi_2} + \frac{\partial\alpha_2}{\partial\xi_1}S - \frac{\partial\alpha_1}{\partial\xi_2}N_1 \tag{2-35}$$

Subsequently, the following two relations can be obtained

$$\frac{\partial\alpha_2 N_2}{\partial\xi_1} - \frac{\partial\alpha_1 S}{\partial\xi_2} - \frac{\partial\alpha_1}{\partial\xi_2}S - \frac{\partial\alpha_2}{\partial\xi_1}N_1 = \alpha_2\frac{\partial N}{\partial\xi_1} - L_1 \tag{2-36}$$

$$\frac{\partial\alpha_1 N_1}{\partial\xi_2} - \frac{\partial\alpha_2 S}{\partial\xi_1} - \frac{\partial\alpha_2}{\partial\xi_1}S - \frac{\partial\alpha_1}{\partial\xi_2}N_2 = \alpha_1\frac{\partial N}{\partial\xi_2} - L_2 \tag{2-37}$$

where

$$N = N_1 + N_2 \tag{2-38}$$

Substituting from (2-34) and (2-36), equation (2-32) can be simplified as

$$\frac{\alpha_1\alpha_2}{R_1}F_1 = i\frac{\varepsilon}{R_1}\left[\alpha_2\frac{\partial N}{\partial\xi_1} - L_1 - \nu\bar{L}_1\right] \tag{2-39}$$

and equation (2-30) becomes

$$L_1 + \frac{\alpha_1\alpha_2}{R_1}F_1 = 0. \tag{2-40}$$

Substituting (2-39) into (2-40) gives

$$L_1 + i\frac{\varepsilon}{R_1}\left[\alpha_2\frac{\partial N}{\partial\xi_1} - L_1 - \nu\bar{L}_1\right] = 0 \tag{2-41}$$

It is easily found that the terms $i\frac{\varepsilon}{R_1}\left(L_1+\nu\bar{L}_1\right)$ is of the order of $\frac{\varepsilon}{R}\sim\frac{h}{R}$ so that is small compared with L_1. Neglecting them yields

$$L_1+i\frac{\varepsilon}{R_1}\alpha_2\frac{\partial N}{\partial \xi_1}=0 \tag{2-42}$$

Similarly, combing (2-31) with (2-33) gives

$$L_2+i\frac{\varepsilon}{R_2}\alpha_1\frac{\partial N}{\partial \xi_2}=0 \tag{2-43}$$

Substituting L_1 from (2-34) and L_2 from (2-35) into (2-42) and (2-43) gives

$$\frac{\partial \alpha_2 N_1}{\partial \xi_1}+\frac{\partial \alpha_1 S}{\partial \xi_2}+\frac{\partial \alpha_1}{\partial \xi_2}S-\frac{\partial \alpha_2}{\partial \xi_1}N_2+i\varepsilon\frac{\alpha_2}{R_1}\frac{\partial N}{\partial \xi_1}=0 \tag{2-44}$$

$$\frac{\partial \alpha_1 N_2}{\partial \xi_2}+\frac{\partial \alpha_2 S}{\partial \xi_1}+\frac{\partial \alpha_2}{\partial \xi_1}S-\frac{\partial \alpha_1}{\partial \xi_2}N_1+i\varepsilon\frac{\alpha_1}{R_2}\frac{\partial N}{\partial \xi_2}=0 \tag{2-45}$$

Subtracting (2-42) and (2-40) gives

$$\alpha_2 F_1=i\varepsilon\frac{\alpha_2}{\alpha_1}\frac{\partial N}{\partial \xi_1} \tag{2-46}$$

A similar relation is

$$\alpha_1 F_2=i\varepsilon\frac{\alpha_1}{\alpha_2}\frac{\partial N}{\partial \xi_2} \tag{2-47}$$

Substituting (2-46) and (2-47) into (2-26) gives

$$\frac{N_1}{R_1}+\frac{N_2}{R_2}-i\varepsilon\frac{1}{\alpha_1\alpha_2}\left[\frac{\partial}{\partial \xi_1}\left(\frac{\alpha_2}{\alpha_1}\frac{\partial N}{\partial \xi_1}\right)+\frac{\partial}{\partial \xi_2}\left(\frac{\alpha_1}{\alpha_2}\frac{\partial N}{\partial \xi_2}\right)\right]=0 \tag{2-48}$$

Equations (2-44), (2-45)and (2-48) are the final equations in terms of “complex forces” called so by Novozhilov (1951).

2.1. Complex Equations for Shells of Revolution

According to (2-38) the complex force N_2 can be expressed by N and N_1, it makes the equations (2-44), (2-45) and (2-48) become

$$\frac{\partial \alpha_2 N_1}{\partial \xi_1}+\frac{\partial \alpha_2}{\partial \xi_1}N_1+\frac{\partial \alpha_1 S}{\partial \xi_2}+\frac{\partial \alpha_1}{\partial \xi_2}S-\frac{\partial \alpha_2}{\partial \xi_1}N+i\varepsilon\frac{\alpha_2}{R_1}\frac{\partial N}{\partial \xi_1}=-\alpha_1\alpha_2 q_1 \tag{2-49}$$

$$-\frac{\partial \alpha_1 N_1}{\partial \xi_2}-\frac{\partial \alpha_1}{\partial \xi_2}N_1+\frac{1}{\alpha_2}\frac{\partial (\alpha_2)^2 S}{\partial \xi_1}+\frac{\partial \alpha_1 N}{\partial \xi_2}+i\varepsilon\frac{\alpha_1}{R_2}\frac{\partial N}{\partial \xi_2}=-\alpha_1\alpha_2 q_2 \tag{2-50}$$

$$\left(\frac{1}{R_1}-\frac{1}{R_2}\right)N_1+\frac{N}{R_2}-i\varepsilon\frac{1}{\alpha_1\alpha_2}\left[\frac{\partial}{\partial \xi_1}\left(\frac{\alpha_2}{\alpha_1}\frac{\partial N}{\partial \xi_1}\right)+\frac{\partial}{\partial \xi_2}\left(\frac{\alpha_1}{\alpha_2}\frac{\partial N}{\partial \xi_2}\right)\right]=q_n' \tag{2-51}$$

where

$$q_n'=q_n+i(1+\nu)\frac{\varepsilon}{\alpha_1\alpha_2}\left(\frac{\partial \alpha_2 q_1}{\partial \alpha_1}+\frac{\partial \alpha_1 q_2}{\partial \alpha_2}\right) \tag{2-52}$$

Note that the loadings q_1, q_2 and q_n are added now.

In the case of the shell of revolution the parameters take the values

$$\xi_1=\theta\,,\ \xi_2=\varphi\,,\ \alpha_1=R_1(\theta),\ \alpha_2=R_2(\theta)\sin\theta \tag{2-53}$$

Substituting (2-53) into (2-49), (2-50) and (2-51) yields

$$\frac{\partial R_2\sin\theta N_1}{\partial \theta}+\frac{\partial R_2\sin\theta}{\partial \theta}N_1+R_1\frac{\partial S}{\partial \varphi}-\frac{\partial R_2\sin\theta}{\partial \theta}N+i\varepsilon\frac{R_2\sin\theta}{R_1}\frac{\partial N}{\partial \theta}=-\alpha_1\alpha_2 q_1 \tag{2-54}$$

$$\frac{\partial (R_2\sin\theta)^2 S}{\partial \theta}-R_1R_2\sin\theta\frac{\partial N_1}{\partial \varphi}+R_1(R_2+i\varepsilon)\sin\theta\frac{\partial N}{\partial \varphi}=-\alpha_1\alpha_2 q_2 \tag{2-55}$$

$$\left(\frac{1}{R_1}-\frac{1}{R_2}\right)N_1+\frac{1}{R_2}N+i\varepsilon\frac{1}{R_1R_2\sin\theta}\left[\frac{\partial}{\partial \theta}\left(\frac{R_2\sin\theta}{R_1}\frac{\partial N}{\partial \theta}\right)+\frac{R_1}{R_2\sin\theta}\frac{\partial}{\partial \varphi}\left(\frac{\partial N}{\partial \varphi}\right)\right]=q_n' \tag{2-56}$$

Novozhilov (1951) introduced two complex variables

$$V=(R_2\sin\theta)^2 S \tag{2-57}$$

$$U = R_2 \sin^2 \theta N_1 - i\varepsilon \frac{R_2 \sin\theta}{R_1} \cos\theta \frac{\partial N}{\partial \theta} \tag{2-58}$$

Then (2-55) can be easily written

$$\frac{\partial V}{\partial \theta} - \frac{R_1}{\sin\theta} \frac{\partial U}{\partial \varphi} - i\varepsilon R_2 \cos\theta \frac{\partial^2 N}{\partial \varphi \partial \theta} + R_1 (R_2 + i\varepsilon) \sin\theta \frac{\partial N}{\partial \varphi} = -R_1 R_2^2 \sin^2 \theta q_2 \tag{2-59}$$

Alternate equation expressed in terms of U , V and N can be obtained by combining (2-53) and (2-55). Multiplying (2-53) in terms of $\sin\theta$ taken together with the Codazzi relation

$$\frac{dR_2 \sin\theta}{d\theta} = R_1 \cos\theta$$

and multiplying (2-56) in terms of $R_1 R_2 \sin\theta \cos\theta$, and adding them yield

$$\frac{\partial U}{\partial \theta} + \frac{R_1}{R_2^2 \sin\theta} \frac{\partial V}{\partial \varphi} - i\varepsilon \frac{R_1 \cos\theta}{R_2 \sin\theta} \frac{\partial^2 N}{\partial \varphi^2} = (q_n' \cos\theta - q_1 \sin\theta) R_1 R_2 \sin\theta \tag{2-60}$$

Eliminating V by a combination of (2-59) and (2-60) gives

$$G(U) - \left\{1 - i\varepsilon \left[\frac{1}{\sin^2\theta}\left(\frac{1}{R_1} - \frac{1}{R_2}\right)\right]\right\} \frac{\partial^2 N}{\partial \varphi^2} = f(\theta, \varphi) \tag{2-61}$$

where the operator G is defined as

$$G = \frac{1}{R_1 R_2 \sin\theta} \frac{\partial}{\partial \theta}\left(\frac{R_2^2 \sin\theta}{R_1} \frac{\partial}{\partial \theta}\right) + \frac{1}{R_2 \sin^2\theta} \frac{\partial^2}{\partial \varphi^2} \tag{2-62}$$

and

$$\begin{aligned} f(\theta,\varphi) = \\ = \frac{1}{R_1 R_2 \sin\theta} \left\{ \frac{\partial}{\partial \theta} \left[(q_n' \cos\theta - q_1 \sin\theta) R_2^3 \sin^2\theta \right] + \frac{\partial q_2}{\partial \varphi} R_1 R_2^2 \sin^2\theta \right\} \end{aligned} \tag{2-63}$$

Multiplying (2-56) in terms of R_2 gives

$$\begin{aligned} \left(\frac{1}{R_1} - \frac{1}{R_2}\right) R_2 N_1 + N + i\varepsilon \frac{1}{R_1 \sin\theta} \frac{\partial}{\partial \theta}\left(\frac{R_2 \sin\theta}{R_1} \frac{\partial N}{\partial \theta}\right) \\ + i\varepsilon \frac{1}{R_2 \sin^2\theta} \frac{\partial^2 N}{\partial \varphi^2} = R_2 q_n \end{aligned} \tag{2-64}$$

According to (2-58) it is easy to verify

$$\left(\frac{1}{R_1}-\frac{1}{R_2}\right)R_2N_1=\left(\frac{1}{R_1}-\frac{1}{R_2}\right)\frac{U}{\sin^2\theta}-i\varepsilon\left(\frac{1}{R_1}-\frac{1}{R_2}\right)\frac{R_2\cos\theta}{R_1\sin\theta}\frac{\partial N}{\partial\theta}$$

Substituting it into (2-64) yields

$$\begin{gathered}\left(\frac{1}{R_1}-\frac{1}{R_2}\right)\frac{U}{\sin^2\theta}+i\varepsilon\left(\frac{1}{R_1}-\frac{1}{R_2}\right)\frac{R_2\cos\theta}{R_1\sin\theta}\frac{\partial N}{\partial\theta}\\ -i\varepsilon\frac{1}{R_1\sin\theta}\frac{\partial}{\partial\theta}\left(\frac{R_2\sin\theta}{R_1}\frac{\partial N}{\partial\theta}\right)+N-i\varepsilon\frac{1}{R_2\sin^2\theta}\frac{\partial^2N}{\partial\varphi^2}=R_2q_n'\end{gathered}\tag{2-65}$$

It is not difficult to verify the equality

$$\begin{aligned}&\frac{1}{R_1\sin\theta}\frac{\partial}{\partial\theta}\left(\frac{R_2\sin\theta}{R_1}\frac{\partial N}{\partial\theta}\right)-\frac{R_2\cos\theta}{R_1\sin\theta}\left(\frac{1}{R_1}-\frac{1}{R_2}\right)\frac{\partial N}{\partial\theta}\\ &\qquad=\frac{1}{R_1R_2\sin\theta}\frac{\partial}{\partial\theta}\left(\frac{R_2^2\sin\theta}{R_1}\frac{\partial N}{\partial\theta}\right)\end{aligned}$$

Substituting it into (2-65) gives

$$\begin{aligned}&\left(\frac{1}{R_1}-\frac{1}{R_2}\right)\frac{U}{\sin^2\theta}+N-i\varepsilon\frac{1}{R_1R_2\sin\theta}\frac{\partial}{\partial\theta}\left(\frac{R_2^2\sin\theta}{R_1}\frac{\partial N}{\partial\theta}\right)-i\varepsilon\frac{1}{R_2\sin^2\theta}\frac{\partial^2N}{\partial\varphi^2}\\ &\qquad=R_2q_n\end{aligned}$$

Namely,

$$-i\varepsilon G(N)+N+\left(\frac{1}{R_1}-\frac{1}{R_2}\right)\frac{U}{\sin^2\theta}=R_2q_n'\tag{2-66}$$

Equations (2-61) and (2-66) are final equations for shells of revolution in terms of the complex variables U and N. The two equations were given by Novozhilov in his monograph in 1951.

2.2. Complex Equations for a Toroidal Shell with Circular Cross Section

For a toroidal shell with circular cross section, as shown in Figure 2.2, the parameters are now

$$R_1 = r_0 \,,\; R_2 = R_0 \sigma / \sin\theta \tag{2-67}$$

where

$$\sigma = 1 + \mu \sin\theta \,,\; \mu = r_0 / R_0 \tag{2-68}$$

Thus, equations (2-61) and (2-66) are rewritten

$$\frac{R_0}{r_0^2 \sigma}\frac{\partial}{\partial\theta}\left(\frac{\sigma^2}{\sin\theta}\frac{\partial U}{\partial\theta}\right) + \frac{1}{R_0\sigma\sin\theta}\frac{\partial^2 U}{\partial\varphi^2} - \left[1 - i\varepsilon\left(\frac{1}{r_0\sigma\sin^2\theta}\right)\right]\frac{\partial^2 N}{\partial\varphi^2} = f(\theta,\varphi) \tag{2-69}$$

and

$$-i\varepsilon\frac{R_0}{r_0^2\sigma}\frac{\partial}{\partial\theta}\left(\frac{\sigma^2}{\sin\theta}\frac{\partial N}{\partial\theta}\right) + \frac{1}{R_0\sigma\sin\theta}\frac{\partial^2 N}{\partial\varphi^2} + N + \frac{1}{r_0\sigma\sin^2\theta}U = \frac{R_0\sigma}{\sin\theta}q_n' \tag{2-70}$$

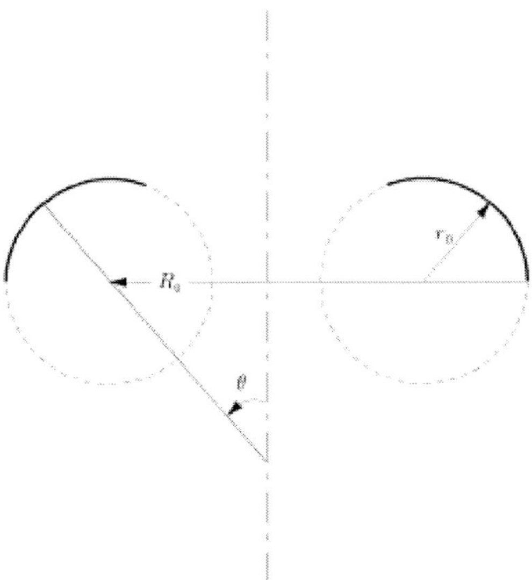

Figure 2.2. Toroidal shells of circular cross section.

where

$$f(\theta,\varphi) = \frac{1}{r_0 R_0 \sigma}\left\{\frac{\partial}{\partial\theta}\left[\left(q_n\cos\theta - q_1\sin\theta\right)\frac{R_0^3\sigma^3}{\sin\theta}\right] + \frac{\partial q_2}{\partial\varphi} r_0 R_0^2 \sigma^2\right\} \tag{2-71}$$

Substituting U from (2-69) into (2-70) gives the final equation in the form

$$
\begin{aligned}
&\sigma^4\frac{\partial^4 N}{\partial\theta^4}+8\mu\sigma^3\cos\theta\frac{\partial^3 N}{\partial\theta^3}\\
&+\left[i\frac{\mu r_0}{\varepsilon}\sigma^3\sin\theta+\sigma^2\left(1+14\mu^2-4\mu\sin\theta-19\mu^2\sin^2\theta\right)\right]\frac{\partial^2 N}{\partial\theta^2}\\
&+\left[i\frac{\mu r_0}{\varepsilon}\sigma^2\cos\theta\left(3+7\mu\sin\theta\right)+\mu\sigma\cos\theta\left(2+4\mu^2-6\mu\sin\theta-12\mu^2\sin^2\theta\right)\right]\frac{\partial N}{\partial\theta}\\
&+\mu^4\frac{\partial^4 N}{\partial\varphi^4}+2\mu^2\sigma^2\frac{\partial^4 N}{\partial\theta^2\partial\varphi^2}+4\mu^3\sigma\cos\theta\frac{\partial^3 N}{\partial\theta\,\partial\varphi^2}\\
&+\left[i\frac{\mu^2 r_0}{\varepsilon}\sigma^2+\mu^3\left(\mu-2\sin\theta-2\mu\sin^2\theta\right)\right]\frac{\partial^2 N}{\partial\varphi^2}\\
&+i\frac{\mu r_0}{\varepsilon}\sigma\left[7\mu+\left(9\mu^2-2\right)\sin\theta-12\mu\sin^2\theta-12\mu^2\sin^3\theta\right]N\\
&=F(\theta,\varphi)
\end{aligned}
\tag{2-72}
$$

where

$$
\begin{aligned}
F(\theta,\varphi)=-i\frac{\mu^2 r_0}{\varepsilon}\sigma\Bigg\{&f(\theta,\varphi)-\frac{r_0}{\mu^2\sigma}\frac{\partial}{\partial\theta}\left[\frac{\sigma^2}{\sin\theta}\frac{\partial}{\partial\theta}\left(q_n'\sigma^2\sin\theta\right)\right]\\
&+\frac{r_0}{\sigma\sin\theta}\frac{\partial^2}{\partial\varphi^2}\left(q_n'\sigma^2\sin\theta\right)\Bigg\}
\end{aligned}
\tag{2-73}
$$

Subsequently, in terms of N, the force and moment resultants N_{11}, $\tilde{N}_{12}$, N_{22}, M_{11}, $\tilde{M}_{12}$ and M_{22} in equations (2-1) – (2-5) can be easily expressed (For simplicity, assume $q_n = \text{constant}$ and neglect q_1 and q_2 here)

$$
N_{11}=-\frac{\varepsilon\sigma}{r_0}\operatorname{Im}\frac{\partial^2 N}{\partial\theta^2}-\frac{\mu\varepsilon}{r_0}\cos\theta\frac{\partial N}{\partial\theta}-\mu\sin\theta\operatorname{Re}N-\frac{\mu^2\varepsilon}{r_0\sigma}\operatorname{Im}\frac{\partial^2 N}{\partial\varphi^2}+q_n r_0\sigma
$$

$$
N_{22}=\frac{\varepsilon\sigma}{r_0}\operatorname{Im}\frac{\partial^2 N}{\partial\theta^2}+\frac{\mu\varepsilon}{r_0}\cos\theta\frac{\partial N}{\partial\theta}+\sigma\operatorname{Re}N+\frac{\mu^2\varepsilon}{r_0\sigma}\operatorname{Im}\frac{\partial^2 N}{\partial\varphi^2}-q_n r_0\sigma
$$

$$
\tilde{N}_{12}=\frac{1}{2}\operatorname{Re}N
$$

$$
\begin{aligned}
M_{11}=&\frac{\varepsilon^2\sigma}{r_0}(1-\nu)\operatorname{Re}\frac{\partial^2 N}{\partial\theta^2}+\frac{\mu\varepsilon^2}{r_0}(1-\nu)\cos\theta\operatorname{Re}\frac{\partial N}{\partial\theta}\\
&+\frac{\mu^2\varepsilon^2}{r_0\sigma}(1-\nu)\operatorname{Re}\frac{\partial^2 N}{\partial\varphi^2}-\varepsilon\left(\sigma-\mu\nu\sin\theta\right)\operatorname{Im}N
\end{aligned}
$$

$$M_{22} = -\frac{\varepsilon^2 \sigma}{r_0}(1-\nu)\mathrm{Re}\frac{\partial^2 N}{\partial \theta^2} - \frac{\mu \varepsilon^2}{r_0}(1-\nu)\cos\theta\,\mathrm{Re}\frac{\partial N}{\partial \theta}$$
$$-\frac{\mu^2 \varepsilon^2}{r_0 \sigma}(1-\nu)\mathrm{Re}\frac{\partial^2 N}{\partial \varphi^2} + \varepsilon(\mu \sin\theta - \nu\sigma)\mathrm{Im}\,N$$

$$\tilde{M}_{12} = -\frac{(1+\nu)\varepsilon}{2}\mathrm{Im}\,N \tag{2-74}$$

Strains $\varepsilon_{11}, \varepsilon_{12}, \varepsilon_{22}$ and curvatures $\kappa_{11}, \kappa_{22}, \tilde{\kappa}_{12}$ can then be obtained by substituting the force and moment resultants into the stress-strain relations (2-21).

3. Toroidal Shells under Non-Axisymmetric Loading

There are few asymptotic solutions to the case of non-axisymmetric loadings. Steele (1959) solved this problem for the first time in his dissertation at Stanford University. He derived a nonhomogeneous integro-differential equation of fourth order and obtained an asymptotic solution for large values of $\mu^* = \sqrt{12(1-\nu^2)}\, r_0^2/(R_0 h)$ and for $m^3 h/\sqrt{12 r_0} \ll 1$. The second condition restricts this solution adaptable for only lower harmonics: $m = 0, 1, 2, \ldots$.

In the present section, a set of asymptotic solutions, including four homogenous solutions and a particular solution, is given in terms of the generally Airy functions introduced by Drazin and Reid (1981) for the fluid stability.

The governing equations for a toroidal shell under non-axisymmetric loadings are (2-72). All the variables and loadings in the equations can be expanded in Fourier series along the circular direction of the shell. The m th order harmonic components are

$$(N,F)(\theta,\varphi) = (N_m, F_m)(\theta)\cos m\varphi \tag{3-1}$$

It implies the associated expansion

$$\begin{aligned}(N_1, N_2, M_1, M_2, q_1, q_n, f, F, u, w)(\theta,\varphi) \\ = (N_{1m}, N_{2m}, M_{1m}, M_{2m}, q_{1m}, q_{nm}, f_m, F, u_m, w_m)(\theta)\cos m\varphi\end{aligned} \tag{3-2}$$

$$(S, H, q_2, v)(\theta,\varphi) = (S_m, H_m, q_{2m}, v_m)(\theta)\sin m\varphi \tag{3-3}$$

Substituting (3-1) into (2-72) gives

$$\begin{aligned}
&\sigma^4\frac{d^4N_m}{d\theta^4}+8\mu\sigma^3\cos\theta\frac{d^3N_m}{d\theta^3}\\
&+\left[i\frac{\mu r_0}{\varepsilon}\sigma^3\sin\theta+\sigma^2\left(1+14\mu^2-2\mu^2m^2-4\mu\sin\theta-19\mu^2\sin^2\theta\right)\right]\frac{d^2N_m}{d\theta^2}\\
&+\left[i\frac{\mu r_0}{\varepsilon}\sigma^2\cos\theta\left(3+7\mu\sin\theta\right)+\mu\sigma\cos\theta\left(2+4\mu^2-4\mu^3m^2-6\mu\sin\theta-12\mu^2\sin^2\theta\right)\right]\frac{dN_m}{d\theta}\\
&+\left\{i\frac{\mu r_0}{\varepsilon}\sigma\left[7\mu-\mu\sigma m^2+\left(9\mu^2-2\right)\sin\theta-12\mu\sin^2\theta-12\mu^2\sin^3\theta\right]\right.\\
&\left.+\mu^3m^2\left(\mu m^2-\mu+2\sin\theta+2\mu\sin^2\theta\right)\right\}N_m=F_m(\theta)
\end{aligned}\tag{3-4}$$

where

$$F_m(\theta)=-i\frac{\mu^2r_0}{\varepsilon}\sigma\left\{f_m(\theta)-\frac{r_0}{\mu^2\sigma}\frac{d}{d\theta}\left[\frac{\sigma^2}{\sin\theta}\frac{d}{d\theta}\left(q_{nm}\sigma^2\sin\theta\right)\right]-m^2r_0\sigma q_{nm}\right\}\tag{3-5}$$

and

$$f_m(\theta)=\frac{1}{r_0R_0\sigma}\left\{\frac{\partial}{\partial\theta}\left[\left(q_{nm}\cos\theta-q_{1m}\sin\theta\right)\frac{R_0^3\sigma^3}{\sin\theta}\right]+mq_{2m}r_0R_0^2\sigma^2\right\}\tag{3-6}$$

In terms of N_m, the m th order harmonic components of stress resultants are expressed

$$\begin{aligned}
N_{1m}=&-\frac{\varepsilon}{r_0}\sigma\,\mathrm{Im}\left(\frac{d^2N_m}{d\theta^2}\right)-\frac{\mu\varepsilon}{r_0}\cos\theta\,\mathrm{Im}\left(\frac{dN_m}{d\theta}\right)-\mu\sin\theta\,\mathrm{Re}\left(N_m\right)\\
&+\frac{\mu^2m^2\varepsilon}{r_0\sigma}\mathrm{Im}\left(N_m\right)+q_{nm}r_0\sigma
\end{aligned}\tag{3-7}$$

$$N_{2m}=\frac{\varepsilon}{r_0}\sigma\,\mathrm{Im}\left(\frac{d^2N_m}{d\theta^2}\right)+\frac{\mu\varepsilon}{r_0}\cos\theta\,\mathrm{Im}\left(\frac{dN_m}{d\theta}\right)+\sigma\,\mathrm{Re}\left(N_m\right)-\frac{\mu^2m^2\varepsilon}{r_0\sigma}\mathrm{Im}\left(N_m\right)-q_{nm}r_0\sigma\tag{3-8}$$

$$\begin{aligned}
mS_m=&\frac{\varepsilon\sigma^2}{\mu r_0}\mathrm{Im}\left(\frac{d^3N_m}{d\theta^3}\right)+4\frac{\varepsilon\sigma}{r_0}\cos\theta\,\mathrm{Im}\left(\frac{d^2N_m}{d\theta^2}\right)+\sigma\sin\theta\,\mathrm{Re}\left(\frac{dN_m}{d\theta}\right)\\
&+\frac{\varepsilon}{\mu r_0}\left(1+2\mu^2-\mu^2m^2-3\mu^2\sin^2\theta\right)\mathrm{Im}\left(\frac{dN_m}{d\theta}\right)+\cos\theta\left(2+3\mu\sin\theta\right)\mathrm{Re}\left(N_m\right)\\
&-\frac{\mu^2m^2\varepsilon}{r_0\sigma}\cos\theta\,\mathrm{Im}\left(N_m\right)-3q_{nm}r_0\sigma\cos\theta
\end{aligned}\tag{3-9}$$

$$\begin{aligned}
M_{1m}=&-\frac{\varepsilon^2}{r_0}(1-\nu)\sigma\,\mathrm{Re}\left(\frac{d^2N_m}{d\theta^2}\right)+\frac{\mu\varepsilon^2}{r_0}(1-\nu)\cos\theta\,\mathrm{Re}\left(\frac{dN_m}{d\theta}\right)\\
&-\frac{\mu^2m^2\varepsilon^2}{r_0\sigma}(1-\nu)\mathrm{Re}\left(N_m\right)-\varepsilon\left(\sigma-\nu\mu\sin\theta\right)\mathrm{Im}\left(N_m\right)-3q_{nm}r_0\sigma\cos\theta
\end{aligned}\tag{3-10}$$

$$M_{2m} = -\frac{\varepsilon^2}{r_0}(1-\nu)\sigma \operatorname{Re}\left(\frac{d^2 N_m}{d\theta^2}\right) - \frac{\mu\varepsilon^2}{r_0}(1-\nu)\cos\theta \operatorname{Re}\left(\frac{dN_m}{d\theta}\right) + \frac{\mu^2 m^2 \varepsilon^2}{r_0\sigma}(1-\nu)\operatorname{Re}(N_m) + \varepsilon(\mu\sin\theta - \nu\sigma)\operatorname{Im}(N_m) \quad (3\text{-}11)$$

$$\begin{aligned} mH_m = &-\frac{\varepsilon^2}{\mu r_0}(1-\nu)\sigma^2 \operatorname{Re}\left(\frac{d^3 N_m}{d\theta^3}\right) - 4\frac{\varepsilon^2}{r_0}(1-\nu)\sigma\cos\theta \operatorname{Re}\left(\frac{d^2 N_m}{d\theta^2}\right) \\ &-\frac{\varepsilon^2}{\mu r_0}(1-\nu)\left(1+2\mu^2-\mu^2 m^2-3\mu^2\sin^2\theta\right)\operatorname{Re}\left(\frac{dN_m}{d\theta}\right) + \varepsilon(1-\nu)\sigma\sin\theta \operatorname{Im}\left(\frac{dN_m}{d\theta}\right) \\ &+\frac{\mu^2 m^2\varepsilon^2}{r_0\sigma}(1-\nu)\cos\theta \operatorname{Re}(N_m) + \varepsilon(1-\nu)\cos\theta(2+3\mu\sin\theta)\operatorname{Im}(N_m) \end{aligned} \quad (3\text{-}12)$$

In above expressions from (3-7) to (3-12), for simplicity, only the m th order normal component of loadings q_{nm} is taken into account.

The m th order harmonic components of displacements u_m and w_m are the respective solutions of the equation

$$\frac{d^2 w_m}{d\theta^2} + w_m = \frac{r_0}{Eh}(N_{1m} - \nu N_{2m}) - \frac{12 r_0^2}{Eh^3}(M_{1m} - \nu M_{2m}) \quad (3\text{-}13)$$

and the equation

$$\frac{du_m}{d\theta} = \frac{r_0}{Eh}(N_{1m} - \nu N_{2m}) - w_m \quad (3\text{-}14)$$

The displacement v_m is

$$\begin{aligned} mv_m = &\frac{r_0\sigma}{\mu Eh}\left\{\frac{\varepsilon\sigma}{r_0}(1+\nu)\operatorname{Im}\left(\frac{d^2 N_m}{d\theta^2}\right) + \frac{\varepsilon\mu}{r_0}(1+\nu)\cos\theta \operatorname{Im}\left(\frac{dN_m}{d\theta}\right)\right. \\ &\left.+\left[1+(1+\nu)\mu\sin\theta\right]\operatorname{Re}(N_m) - \frac{\mu^2 m^2\varepsilon}{r_0\sigma}(1+\nu)\operatorname{Im}(N_m) - q_{nm}(1+\nu)r_0\sigma\right\} \\ &-(u_m\cos\theta + w_m\sin\theta) \end{aligned} \quad (3\text{-}15)$$

In order to find such an asymptotic solution of equation (2-72) that is valid uniformly in the entire toroidal, referring to the Langer transformation, the following transformation is defined

$$z = \left[\frac{3}{2m^2}\int_0^\theta \left(\frac{\sin\theta}{1+\mu\sin\theta}\right)^{1/2} d\theta\right]^{2/3} \quad (3\text{-}16)$$

and

$$N_m = \frac{m^3}{(1+\mu\sin\theta)^2}\left(\frac{\sin\theta}{1+\mu\sin\theta}\right)^{-3/4}\psi \tag{3-17}$$

Thus, equation (2-72) is transformed to

$$\frac{d^4\psi}{dz^4}+\lambda^2\left(z\frac{d^2\psi}{dz^2}+L_1\frac{d\psi}{dz}+L_2\psi\right)+\left(L_3\frac{d^2\psi}{dz^2}+L_4\frac{d\psi}{dz}+L_5\psi\right)=\lambda^2 G_n+G_\tau \tag{3-18}$$

in which λ is a large complex-valued parameter

$$\lambda^2 = i\sqrt{12\left(1-\nu^2\right)}\frac{\mu r_0}{h}m^4 \tag{3-19}$$

$$L_1 = \frac{1}{m^4\sigma}\frac{1}{\phi_0 z'^4}\left[\left(\phi_0 z'' + 2\phi_0' z'\right)\sin\theta + 3\phi_0 z'\cos\theta\right]$$

$$L_2 = \frac{1}{m^4\sigma}\frac{1}{z'^4}\left\{\frac{1}{\phi_0}\left(\phi_0''\sin\theta+3\phi_0'\cos\theta\right)\right.$$
$$\left.+\frac{1}{\sigma^2}\left[\mu\left(1-m^2\right)+\left(\mu^2-\mu^2m^2-2\right)\sin\theta-4\mu\sin^2\theta-2\mu^2\sin^3\theta\right]\right\}$$

$$L_3 = \frac{1}{\sigma^2}\frac{1}{z'^2}\left(1+2\mu^2-\mu^2m^2+8\mu\sin\theta+5\mu^2\sin^2\theta\right)$$
$$+\frac{1}{\phi_0 z'^4}\left(4\phi_0 z'z'''+3\phi_0 z''^2+12\phi_0' z'z''+6\phi_0'' z'^2\right)$$

$$L_4 = \frac{1}{\phi_0 z'^4}\left[\left(\phi_0 z^{(4)}+4\phi_0' z'''+6\phi_0'' z''+4\phi_0''' z'\right)\right.$$
$$\left.+\frac{1}{\sigma^2}\left(\phi_0 z''+2\phi_0' z'\right)\left(1+2\mu^2-2\mu^2m^2+8\mu\sin\theta+5\mu^2\sin^2\theta\right)\right]$$
$$+\frac{\mu\cos\theta}{\sigma^2 z'^3}\left(6-4\mu^2+4\mu^2m^2+2\mu\sin\theta\right)$$

$$L_5 = \frac{\phi_0^{(4)}}{\phi_0 z'^4}+\frac{\phi_0''}{\sigma^2\phi_0 z'^4}\left(1+2\mu^2-2\mu^2m^2+8\mu\sin\theta+5\mu^2\sin^2\theta\right)$$
$$+\frac{\phi_0'\mu\cos\theta}{\sigma^3\phi_0 z'^4}\left(6-4\mu^2+4\mu^2m^2+2\mu\sin\theta\right)$$
$$-\frac{\mu^2}{\sigma^4 z'^4}\left[\left(6-4\mu^2-\mu^2m^4+5\mu^2m^2\right)+2\mu m^2\sin\theta\right.$$
$$\left.-\left(12-2\mu^2+2\mu^2m^2\right)\sin^2\theta-12\mu\sin^3\theta-4\mu^2\sin^4\theta\right]$$

and

$$G_n = -\frac{r_0 m}{\mu\sigma^2}\left(\frac{z\sigma}{\sin\theta}\right)^{\frac{5}{4}} q_{mm}\left\{\mu^2 m^2\left(1+\frac{\mu}{2}\right)+\left[\left(2m^2+\frac{9}{4}\right)\mu^3+3\mu\right]\sin\theta\right.$$
$$\left.-\left[\frac{1}{2}\left(m^2+3\right)\mu^4+9\mu^2\right)\right]\cos 2\theta-\frac{27}{4}\mu^3\sin 3\theta+\frac{3}{2}\mu^4\sin 4\theta\right\}$$

where

$$\phi_0 = m^3\left(\frac{\sin\theta}{\sigma z}\right)^{-3/4}$$

and

$$(\,)' = \frac{d}{d\theta}(\,)$$

The expression of G_τ is omitted.

L_1, L_2, L_3, L_4, L_5, G_n and G_τ are real-valued functions of z and are analytic at the transition point $z=0$. The non-homogeneous term G_τ vanishes if the loadings have only normal components. Particularly, it is necessary to indicate that

$$L_1(0)=3 \tag{3-20}$$

Equation (3-18) is the fundamental equation for asymptotic analysis.

3.1. Local Solutions

In order to assume the form of the asymptotic expansion, we find firstly a "local solutions" which are valid only in the vanity of the transition point $z=0$. It is evident that the local solutions can be obtained by letting $z\to 0$ in the solutions which are valid everywhere in the entire toroidal. Therefore, they can help us infer the extensive forms of the uniformly valid solutions.

In order to find the local solutions we derive the "local equation" which is valid also only in the vanity of the transition point $z=0$. Letting

$$\zeta = -\lambda^{2/3} z \tag{3-21}$$

and using the constants

$$l_2 = L_2(0), l_3 = L_3(0), l_4 = L_4(0), g_n = G_n(0) \tag{3-22}$$

instead of $L_1(z)$, $L_2(z)$, $L_3(z)$, $L_4(z)$ and $G_n(z)$, equation (3-18) is transferred to

$$\frac{d^4\psi}{d\zeta^4} - \zeta\frac{d^2\psi}{d\zeta^2} - 3\frac{d\psi}{d\zeta} + \lambda^{-2/3}l_2\psi + \lambda^{-4/3}l_3\frac{d^2\psi}{d\zeta^2} - \lambda^{-6/3}l_4\frac{d\psi}{d\zeta} = \lambda^{-2/3}g_n \tag{3-23}$$

in which (3-20) is used and the small terms of order $O\left(\lambda^{-8/3}\right)$ are neglected because they have exceed the accuracy of the thin shell theory.

3.1.1. Comparison Equation

Expand ψ into the following form

$$\psi(\zeta, \lambda) = \sum_0^\infty \left(\lambda^{-2/3}\right)^n \psi_n(\zeta) \tag{3-24}$$

Substituting (3-24) into the homogeneous part of (3-23) and equating coefficients of like powers of $\lambda^{-2/3}$, we obtain

$$\boldsymbol{TD}\psi_0 = 0 \tag{3-25}$$

$$\boldsymbol{TD}\psi_n = -l_2\psi_{n-1} - l_3\frac{\partial^2\psi_{n-2}}{\partial\zeta^2} + l_4\frac{\partial\psi_{n-3}}{\partial\zeta} \quad (n = 1,2,3\ldots) \tag{3-26}$$

where such the coefficients are defined to be zero when their subscripts are negative. The differential operator $\boldsymbol{T}$ is defined as

$$\boldsymbol{T} = \boldsymbol{D}^3 - \zeta\boldsymbol{D} - 3, \ \boldsymbol{D} = \frac{d}{d\zeta} \tag{3-27}$$

The fourth order equation (3-25) is referred to as a comparison equation. Solutions of the comparison equation will be used as a set of basic functions to expand the solutions of the original equation (3-4).

3.1.2. Asymptotic Expansions of the First Two Homogenous Solutions

Now we give the four homogeneous solutions to the comparison equation (3-25).

It is seen by putting $p = -2$ in (A7) and (A8) and $p = 0$ in (A9), that $A_k(\zeta, -1)$, $B_k(\zeta, -1)$ and $B_0(\zeta, 1)$ $(k = 1,2,3)$ are non-trivial solutions of the homogenous equation (3-25). Moreover, we indicate that $A_2(\zeta, -1), A_3(\zeta, -1)$, $B_1(\zeta, -1)$ and

$B_0(\zeta,1)\equiv 1$ are a linearly independent set of solutions, because their Wronskian, which can be derived from (A17), is a constant and is given by

$$\mathcal{W}(A_2,A_3,B_1,1)(\zeta,-1)=-\frac{i}{\pi} \tag{3-28}$$

So that the four homogeneous solutions of the comparison equation (3-25) are

$$\psi_0^{(k)}=A_{k+1}(\zeta,-1),\ \psi_0^{(3)}=B_1(\zeta,-1),\ \psi_0^{(4)}=B_0(\zeta,1)=1 \tag{3-29}$$

where $k=1,2$.

The higher order approximation can be found by substituting (3-29) into (3-26). For example, substituting $\psi_0^{(k)}$ into (3-26) with $n=1$ and comparing with (A7) with $p=-1$, we obtain

$$\psi_1^{(k)}=l_2A_{k+1}(\zeta,0) \tag{3-30}$$

Then, substituting $\psi_0^{(k)}$ and $\psi_1^{(k)}$ into (3-26) with $n=2$, noting $(A5)_1$ and comparing with (A7) with $p=0$ and $p=-3$, respectively, we obtain

$$\psi_2^{(k)}=\frac{1}{2}l_2^2A_{k+1}(\zeta,1)-l_3A_{k+1}(\zeta,-2,0) \tag{3-31}$$

where $A_{k+1}(\zeta,-2,0)$ is used according to (A23), instead of $A_{k+1}(\zeta,-2)$.

Moreover, substituting $\psi_0^{(k)}$, $\psi_1^{(k)}$ and $\psi_2^{(k)}$ into (3-26) with $n=3$, noting $(A5)_1$, using (A7) with $p=1$ and using (A21) with $p=-2$ and $q=1$, respectively, we obtain

$$\psi_3^{(k)}=\frac{1}{6}l_2^3A_{k+1}(\zeta,2)+l_4A_{k+1}(\zeta,-1,1) \tag{3-32}$$

Similarly, $\psi_4^{(k)}$, $\psi_5^{(k)}$,... can then be obtained by a repeat in the same way. Substituting all of them into (3-24) and using (A11) and (A22), we conclude that the formal expansions are of the form

$$\begin{aligned}\psi^{(k)}(\zeta,\lambda)&=\alpha A_{k+1}(\zeta,-1)+\lambda^{-2/3}\beta A_{k+1}(\zeta,0)+\left(\lambda^{-2/3}\right)^2\gamma A_{k+1}(\zeta,1)\\&+\lambda^{-2}\left\{aA_{k+1}(\zeta,-1,1)+\lambda^{-2/3}bA_{k+1}(\zeta,0,1)+\left(\lambda^{-2/3}\right)^2cA_{k+1}(\zeta,1,1)\right\}\\&+\ldots\end{aligned} \tag{3-33}$$

where the coefficients α, β... are the constants composed of l_2, l_3 and l_4, and have the expansions for λ^{-2} like that

$$\alpha=\alpha(\lambda)=\sum_{n=0}^{\infty}\alpha_n\left(\lambda^{-2}\right)^n,\ \beta=\beta(\lambda)=\sum_{n=0}^{\infty}\beta_n\left(\lambda^{-2}\right)^n \tag{3-34}$$

3.1.3. Asymptotic Expansions of the Third Homogenous Solution

In a similar way, the formal expansion of the third independent solution of (3-23), whose first approximation is $\psi_0^{(3)}$ as indicated in (3-29), is of the form

$$\begin{aligned}\psi^{(3)}(\zeta,\lambda)&=\alpha B_1(\zeta,-1)+\lambda^{-2/3}\beta B_1(\zeta,0)+\left(\lambda^{-2/3}\right)^2\gamma B_1(\zeta,1)+\left(\lambda^{-2/3}\right)^2\delta\\&+\lambda^{-2}\left\{aB_1(\zeta,-1,2)+\lambda^{-2/3}bB_1(\zeta,0,2)+\left(\lambda^{-2/3}\right)^2cB_1(\zeta,1,2)\right\}\\&+\ldots\end{aligned} \tag{3-35}$$

in which the coefficients, except δ, are denoted in terms of the same notations as in (3-33) because they are equal to each other.

3.1.4. Asymptotic Expansions of the Fourth Homogenous Solution

In order to derive the formal expansion of the fourth independent solution to (3-23), whose first approximation is $\psi_0^{(4)}=1$ as indicated in (3-29), we begin with the non-homogeneous equation of the form

$$\boldsymbol{TD}\psi_1^{(4)}=-l_2 \tag{3-36}$$

which is derived out by a substitution of $\psi_0^{(4)}=1$ into (3-26) with $n=1$.

It is easily seen, let $p=1$ in (A5)$_3$ and note (A9), that $-\frac{1}{3}B_0(\zeta,2)$ is a particular solution to the non-homogeneous equation $\boldsymbol{TD}u=1$. Thus,

$$\psi_1^{(4)}=\frac{1}{3}l_2B_0(\zeta,2) \tag{3-37}$$

Subsequently, we can find $\psi_2^{(k)}$, $\psi_3^{(k)}$... which are expressed in terms of $B_0(\zeta,p)$ with $p=1,2,3\ldots$. Note from (A10) that $B_0(\zeta,p)$ is a polynomial in ζ. Then, by noting (3-21) and (3-24) we know that the complete expansion of the fourth solution is of the form

$$\psi^{(4)}=\psi^{(4)}(z,\lambda)=\sum_{n=0}^{\infty}\left(\lambda^{-2}\right)^n\psi_n^{(4)}(z) \tag{3-38}$$

Note that $\psi^{(4)}$ is a function of z instead of ζ, which means that $\psi^{(4)}$ is a slowly varying function.

3.1.5. Asymptotic Expansion of a Particular Solution

A particular solution of (3-23) is assumed having the form:

$$\psi^*(\zeta,\ \lambda)=\lambda^{-2/3}\sum_{n=0}^{\infty}\left(\lambda^{-2/3}\right)^n\psi_n^*(\zeta) \tag{3-39}$$

A substitution of (3-39) into (3-23) yields a series of equations in the form:

$$\boldsymbol{TD}\psi_0^* = g_n \tag{3-40}$$

$$\boldsymbol{TD}\psi_n^* = -l_2\psi_{n-1}^* - l_3\frac{\partial^2\psi_{n-2}^*}{\partial\zeta^2} + l_4\frac{\partial\psi_{n-3}^*}{\partial\zeta} \quad (n=1,2,3...) \tag{3-41}$$

It is found that (3-40) is similar to (3-36) and (3-41) is similar to (3-26). So that the particular solution ψ^* has a similar expansion to $\psi^{(4)}$ in the form

$$\psi^* = \psi^*(z,\lambda)=\sum_{n=0}^{\infty}\left(\lambda^{-2}\right)^n\psi_n^*(z) \tag{3-42}$$

Similarly, ψ^* is a slowly varying function.

This is totally different from the situation in the case of axisymmetric loadings where the particular solution is a rapidly varying function. Please see the fourth section of the paper.

As mentioned above that all of these solutions are valid only in the vicinity of the transition point. However, we may think of that they are also uniformly valid in the entire toroid if the coefficients in them are functions of z instead of constants. In other words, we will find such the uniformly valid solutions of (3-18) that they have the same expressions as (3-33), (3-35), (3-38) and (3-42), in which, however, the coefficients are an unknown function of z.

3.2. Uniformly Valid Expansions in an Entire Toroidal Shell

Now we assume that the coefficients in (3-33), (3-35), (3-38) and (3-42) are functions of z instead of constants, for example,

$$\alpha=\alpha(z,\lambda)=\sum_{n=0}^{\infty}\alpha_n(z)\left(\lambda^{-2}\right)^n,\ \beta=\beta(z,\lambda)=\sum_{n=0}^{\infty}\beta_n(z)\left(\lambda^{-2}\right)^n \tag{3-43}$$

3.2.1. The First and Second Homogeneous Solution

Substituting (3-33) into (3-18), using (A5), (A11), (A20), (A22) and (A23), then equating the coefficients of $A_{k+1}(\zeta, p, q)$ $(p=-1,2,1; q=0,1,2...)$, we obtain a set of ordinary differential equations in terms of the unknown coefficients in (3-33). Writing these coefficients in the form of the asymptotic series as (3-43) gives

$$2z^2\alpha_0' + 3z\alpha_0 - L_1\alpha_0 = 0 \tag{3-44$_1$}$$

$$2z^2\beta_0' - (L_1 - 2)\beta_0 = 5z\alpha_0'' + 8\alpha_0' - L_1\alpha_0' - L_2\alpha_0 + zL_3\alpha_0 \tag{3-44$_2$}$$

$$z\gamma_0'' + L_1\gamma_0' + L_2\gamma_0 = 0 \tag{3-44$_3$}$$

where the prime indicates the differentiation with respect to z. The solution of these equations gives a complete asymptotic expansion within the accuracy of the thin shell theory in the form:

$$\begin{aligned}\psi^{(k)}(\zeta,\lambda) &= \alpha_0 A_{k+1}(\zeta,-1) + \lambda^{-2/3}\beta_0 A_{k+1}(\zeta,0) + \lambda^{-4/3}\gamma_0 A_{k+1}(\zeta,1) \\ &+ \lambda^{-2} a_0 A_{k+1}(\zeta,-1,1) \qquad (k=1,2)\end{aligned} \tag{3-45}$$

where

$$\alpha_0(z) = a_0(z) = z^{-3/2}\exp\left\{\frac{1}{2}\int_0^z \frac{L_1(\tau)}{\tau} d\tau\right\} \tag{3-46}$$

$$\beta_0(z) = \frac{1}{8}z^{-1}\exp\left\{\frac{1}{2}\int_0^z \frac{L_1(\tau)}{\tau} d\tau\right\}\int_0^z M(\tau)d\tau \tag{3-47}$$

$$M(\tau) = \tau^{-5/2}\left[27 - 18L_1(\tau) + 3L_1^2(\tau)\right] + 2\tau^{-3/2}\left[5L_1'(\tau) - 2L_2(\tau)\right] + 4L_3(\tau) \tag{3-48}$$

It is argued that $\alpha_0(z)$ and $\beta_0(z)$ are analytic at the transition point $z = 0$.

In order to find $\gamma_0(z)$, which is a solution of equation (3-44)$_3$, we indicate that (3-44)$_3$ is the homogeneous counterpart of the membrane equation

$$z\phi'' + L_1\phi' + L_2\phi = G_n \tag{3-49}$$

which is obtained by assuming $\lambda \to \infty$ in the fundamental equation (3-18). It can be shown that only one homogeneous solution of (3-49), specified by $\phi_1^{(0)}(z)$, is analytic at the transition point $z = 0$ and the other independent homogeneous solution of (3-49), specified by $\phi_2^{(0)}(z)$, is singular at $z=0$. Moreover, there exists such a particular solution of (3-49), specified by $\phi_*^{(0)}(z)$, which is analytic at $z=0$. So that, we let

$$\gamma_0(z)=\phi_1^{(0)}(z) \tag{3-50}$$

3.2.2. The Third Homogeneous Solution

Similarly, substituting (3-35) into (3-18) gives a complete expansion of the third homogeneous solution within the accuracy of the thin shell theory in the form

$$\psi^{(3)}(\zeta,\lambda)=\alpha_0 B_1(\zeta,-1)+\lambda^{-2/3}\beta_0 B_1(\zeta,0)+\lambda^{-4/3}\gamma_0 B_1(\zeta,1,1)+\lambda^{-4/3}\delta_0(z) \tag{3-51}$$

Note that the last term in (3-51) is a slowly varying function. Coefficients $\alpha_0(z)$, $\beta_0(z)$ and $\gamma_0(z)$ are determined still by equations (3-46), (3-47) and (3-50), respectively. The slowly varying function $\delta_0(z)$ is a solution to the non-homogeneous equation

$$z\delta_0''+L_1\delta_0'+L_2\delta_0=\frac{x(z)}{z} \tag{3-52}$$

where

$$x(z)=-6\alpha_0''-L_3\alpha_0-z\beta_0''+4\beta_0'-L_1\beta_0'-L_2\beta_0+2z\gamma_0'+L_1\gamma_0-\gamma_0 \tag{3-53}$$

Thus,

$$\delta_0(z)=\int_0^z\frac{\phi_1^{(0)}(\tau)\phi_2^{(0)}(z)-\phi_1^{(0)}(z)\phi_2^{(0)}(\tau)}{\phi_1^{(0)}(\tau)\phi_2^{(0)\prime}(\tau)-\phi_1^{(0)\prime}(\tau)\phi_2^{(0)}(\tau)}\frac{x(\tau)}{\tau}d\tau \tag{3-54}$$

3.2.3. The Fourth Homogeneous Solution and the Particular Solution

It is sufficient to hold the leading term in the expansions (3-38) and (3-42) in powers of λ^{-2}. Substituting (3-38) and (3-42) into (3-18), we find that the leading terms of ψ_0^* and $\psi_0^{(4)}$ satisfy the membrane equation (3-49) and its homogeneous counterpart, respectively. Therefore, we let

$$\psi^{(4)}\approx\psi_0^{(4)}=\phi_1^{(0)}(z) \tag{3-55}$$

and

$$\psi^* \approx \psi_0 = \phi_*^{(0)}(z) \tag{3-56}$$

Of course, all of them are slowly varying functions.

3.3. Final Results

A general solution of the fundamental equation (3-18) is of the form

$$\psi(\zeta,\lambda) = C_1\psi^{(1)} + C_2\psi^{(2)} + C_3\psi^{(3)} + C_4\psi^{(4)} + \psi^* \tag{3-57}$$

where the homogeneous solutions $\psi^{(1)}$, $\psi^{(2)}$ $\psi^{(3)}$ and $\psi^{(4)}$ as well as the particular solution ψ^* are given in (3-45), (3-51), (3-53) and (3-54), respectively. Four complex constants C_i $(i = 1,2,3,4)$ will be determined according to eight boundary conditions.

The next step is to find the complex variable N_m according to (3-17). Then the stress resultants and displacements can be determined in terms of equations from (3-7) to (3-15).

3.4. "Wind-Type" Loadings: Comparison with Novozhilov's Results

Let $m = 1$ in (3-2) and (3-3), the loadings become

$$q_1 = q_{11}(\theta)\cos\varphi,\ q_2 = q_{21}(\theta)\sin\varphi,\ q_n = q_{n1}(\theta)\cos\varphi \tag{3-58}$$

which is referred to as "wind-type" loadings by Novozhilov (1951).

Novozhilov (1951) in his monograph gave an asymptotic expression of the force and moment resultants and the strains for the general shell of revolution subjected to the wind-type loading based on the limit that the special circumferences $\theta = 0$ and $\theta = \pi$ are not on the shell. Obviously, his expressions are still valid for a segment of toroidal shell with positive curvature between and away from $\theta = 0$ and $\theta = \pi$.

In the other respects, the results in the present paper are adaptable for the toroidal shell with any non-axisymmetric loadings. Their special case of $m = 1$ and $\theta \neq 0$ and $\theta \neq \pi$ of course must coincide with Novozhilov's results.

Novozhilov's results are only first approximation. All the small quantities higher than order $O(\sqrt{\varepsilon}) \sim O(\sqrt{h})$ in them are omitted. With the same accuracy formula (3-57) becomes

$$\psi(\zeta,\lambda) = \alpha_0(z)[C_1A_2(\zeta,-1) + C_2A_3(\zeta,-1)] + C_4\phi_1^{(0)}(z) + \phi_*^{(0)}(z) \tag{3-59}$$

in which, according to the formulae from (A12) to (A17), A_2 and A_3 are expressed as

$$A_2(\zeta,-1)\approx \tilde{f}(\theta,\varepsilon)[ie^{(1-i)\omega}+e^{-(1-i)\omega}] \tag{3-60$_1$}$$

$$A_3(\zeta,-1)\approx \tilde{f}(\theta,\varepsilon)[-ie^{(1-i)\omega}+e^{-(1-i)\omega}] \tag{3-60$_2$}$$

where

$$\tilde{f}(\theta,\varepsilon)=\varepsilon^{-1/12}(-i\mu r_0)^{1/12}z^{1/4} \tag{3-61}$$

$$\omega=\frac{1}{\sqrt{2}}\int_{\theta_0}^{\theta}\sqrt{\frac{\sin\theta}{R_0\sigma\varepsilon}}r_0 d\theta \tag{3-62}$$

and $\theta_0\neq 0$ is the upper boundary circumference.

Substituting (3-59)-(3-62) into (3-59) and (3-17) yields

$$\begin{aligned} N_1 &= \frac{1}{\sigma^2}\left(\frac{\sin\theta}{\sigma z}\right)^{-3/4}\tilde{f}(\theta,\varepsilon)\alpha_0(z)\left[\tilde{C}_1 e^{-(1-i)\omega}+\tilde{C}_2 e^{(1-i)\omega}\right] \\ &+\frac{1}{\sigma^2}\left(\frac{\sin\theta}{\sigma z}\right)^{-3/4}\left[C_4\phi_1^{(0)}(z)+\phi_*^{(0)}(z)\right] \end{aligned} \tag{3-63}$$

where

$$\widetilde{C}_1=C_1+C_2,\ \widetilde{C}_2=i(C_1-C_2) \tag{3-64}$$

In the first term of (3-63), the coefficient $\frac{1}{\sigma^2}\left(\frac{\sin\theta}{\sigma z}\right)^{-3/4}\tilde{f}(\theta,\varepsilon)\alpha_0(z)$ has to be seen as a constant, since only the primary term is retained in the derivative of $\widetilde{T}_1$ with respect to θ . It can be absorbed by the arbitrary complex constants $\widetilde{C}_1$ and $\widetilde{C}_2$. Thus, (3-63) is rewritten as

$$N_1=\tilde{\tau}+N^{(0)} \tag{3-65}$$

in which

$$\tilde{\tau}=\widetilde{C}_1 e^{-(1-i)\omega}+\widetilde{C}_2 e^{(1-i)\omega} \tag{3-66}$$

and

$$N^{(0)}=\frac{1}{\sigma^2}\left(\frac{\sin\theta}{\sigma z}\right)^{-3/4}\left[C_4\phi_1^{(0)}(z)+\phi_*^{(0)}(z)\right] \tag{3-67}$$

It can be seen that (3-66) is the same as formula (4.13.24) in Novozhilov's monograph and that (3-67) satisfies the membrane equation of (3-4) or the membrane equation of (2-49)-(2-51) with (2-52) or the membrane equation of (3-4) with $m=1$. The reason of the latter is because $C_4\phi_1^{(0)}+\phi_*^{(0)}$ in (3-67) satisfies the membrane equation (3-49) and because of (3-17).

All the stress resultants and stress couples can be obtained by substituting (3-65) into (3-7) - (3-15). As an example we consider

$$\begin{aligned}N_{11}&=-\frac{\mu\varepsilon}{r_0}\cos\theta\,\mathrm{Im}\left(\frac{d\tilde{\tau}}{d\theta}\right)\\&+\left\{-\frac{\varepsilon\sigma}{r_0}\mathrm{Im}\left(\frac{d^2\tilde{\tau}}{d\theta^2}\right)-\mu\sin\theta\,\mathrm{Re}(\tilde{\tau})-\mu\sin\theta\,\mathrm{Re}(T^{(0)})q_{n1}r_0\sigma\right\}\end{aligned} \tag{3-68}$$

where

$$\begin{aligned}&-\frac{\mu\varepsilon}{r_0}\cos\theta\,\mathrm{Im}\left(\frac{d\tilde{\tau}}{d\theta}\right)\\&=-\frac{1}{\sqrt{2}}\sqrt{\frac{\varepsilon\sin\theta}{R_0\sigma}}\cos\theta\left\{\left[(C_1'-C_1'')\cos\omega-(C_1'+C'')\sin\omega\right]e^{-\omega}\right.\\&\qquad\left.-\left[(C_2'-C_2'')\cos\omega+(C_2'+C_2'')\sin\omega\right]e^{\omega}\right\}\end{aligned} \tag{3-69}$$

It can be seen that (3-69) is identical to Novozhilov's formula (3.14.9) if $\cos\theta$ is replaced with $\cot\theta$. However, both $\cos\theta$ and $\cot\theta$ can be absorbed in the arbitrary constants C_1', C_1'', C_2' and C_2'' as mentioned above. Thus, we can see (3-69) is as the same as Novozhilov's resultant.

In addition, the second summand of (3-68) is equal to the membrane stress resultant $N_{11}^{(0)}$. The real part of the complex constant C_4 in $N_{11}^{(0)}$ [see (3-67)] can be determined by giving the boundary membrane stress resultant

$$N_{11}^*=N_{11}^{(0)}\Big|_{\theta=\theta_0}=\left\{-\frac{\varepsilon\sigma}{r_0}\mathrm{Im}\left(\frac{d^2\tilde{\tau}}{d\theta^2}\right)-\mu\sin\theta\,\mathrm{Re}(\tilde{\tau})-\mu\sin\theta\,\mathrm{Re}(T^{(0)})+q_{n1}r_0\sigma\right\}\Bigg|_{\theta=\theta_0} \tag{3-70}$$

As for the determination of the imaginary part of C_4, we have to give the boundary shear stress resultant S^*. This is the same as Novozhilov's approach.

Finally, we have

$$N_{11} = -\frac{1}{\sqrt{2}}\sqrt{\frac{\varepsilon \sin\theta}{R_0\sigma}}\cos\theta\{[(C_1' - C_1'')\cos\omega - (C_1' + C'')\sin\omega]e^{-\omega} - [(C_2' - C_2'')\cos\omega + (C_2' + C_2'')\sin\omega]e^{\omega}\} + N_{11}^{(0)} \tag{3-71}$$

which is identical to Novozhilov's formulae (3.14.9) or (3.14.22).

Note that there are only four constants C_1', C_1'', C_2' and C_2'' to be determined in the present case instead of eight constants in the other non-axisymmetric cases. This is a special feature of the wind-type loading on which Novozhilov has made a detailed explanation.

4. Axisymmetric Deformation of a Toroidal Shell with Circular Cross Section

A toroidal shell suffers an axisymmetric deformation when both the surface loading and boundary loading are axisymmetric. In this case, as mentioned before, let $m = 0$ in all the m th order harmonic components given in (3-1) - (3-3). Particularly, the surface loadings become

$$q_1 = q_{11}(\theta),\ q_2 = q_{21}(\theta) = 0,\ q_n = q_{n1}(\theta) \tag{4-1}$$

In this section we firstly derive an axisymmetric equation for shells of revolution and then give an axisymmetric equation for a toroidal shell with circular cross section.

The paragraph 4.1 and 4.2 are cited from Novozhilov's monograph in 1951.

4.1. Axisymmetric Equation for Shells of Revolution

We begin with the fundamental equations for shells of revolution (2-61) and (2-66). Equation (2-61) comes from (2-59) and (2-60). However (2-59) becomes an identity when $m = 0$, so that equation (2-61) has to be replaced by (2-60). Moreover, when $m = 0$, equation (2-60) is reduced to the form

$$\frac{dU}{d\theta} = (q_n \cos\theta - q_1 \sin\theta) R_1 R_2 \sin\theta \tag{4-2}$$

An integration of (4-2) is easily obtained in the form:

$$U = \int_{\theta_0}^{\theta} (q_n \cos\theta - q_1 \sin\theta) R_1 R_2 \sin\theta d\theta + C_1 \tag{4-3}$$

where C_1 is a real integral constant.

Substituting (4-3) into (2-66) gives

$$\frac{d^2N}{d\theta^2}+\left[\left(2\frac{R_1}{R_2}-1\right)\frac{\cos\theta}{\sin\theta}-\frac{1}{R_1}\frac{dR_1}{d\theta}\right]\frac{dN}{d\theta}+i\frac{R_1^2}{\varepsilon R_2}N=i\frac{R_1^2}{\varepsilon R_2}F(\theta) \tag{4-4}$$

where

$$F(\theta)=R_2q_n-\left(\frac{1}{R_1}-\frac{1}{R_2}\right)\frac{1}{\sin^2\theta}\left[\int_{\theta_0}^{\theta}(q_n\cos\theta-q_1\sin\theta)R_1R_2\sin\theta d\theta+C_1\right] \tag{4-5}$$

Equation (4-4) is the governing equation for shells of revolution in an axisymmetric deformation.

4.2. Axisymmetric Equation for a Toroidal Shell with Circular Cross Section

For a toroidal shell with circular cross section the geometrical parameters take the values as given in (2-67) and (2-68). A substitution of these parameters into (4-4) and (4-5) gives

$$\begin{aligned}&\frac{1}{1+\mu\sin\theta}\frac{d}{d\theta}\left[\frac{(1+\mu\sin\theta)^2}{\sin\theta}\frac{dN}{d\theta}\right]+i\kappa N\\&=i\kappa\left[\frac{pr_0}{2}\frac{3+2\mu\sin\theta}{1+\mu\sin\theta}+\frac{A}{(1+\mu\sin\theta)\sin^2\theta}\right]\end{aligned} \tag{4-6}$$

where

$$\kappa=\frac{r_0^2}{\varepsilon R_0}=\frac{\sqrt{12(1-\nu^2)}r_0^2}{hR_0} \tag{4-7}$$

is a large real number. In (4-6), only a uniform normal pressure as the loading is considered. A is a real constant.

Obviously, coefficients in (4-6) approach to infinity at the transition points of $\theta=0$ and $\theta=\pm\pi$. Therefore, it is in need of some modification. Differentiating (4-6) with respect to θ leads to an improved equation in the form

$$\begin{aligned}&(1+\mu\sin\theta)^2\frac{d}{d\theta}\left[\frac{1}{1+\mu\sin\theta}\frac{dZ}{d\theta}\right]+i\kappa\sin\theta Z\\&=-i\kappa\cos\theta\left[\mu\frac{pr_0}{2}+A\frac{2+3\mu\sin\theta}{\sin^2\theta}\right]\end{aligned} \tag{4-8}$$

in which an auxiliary function is defined as

$$Z = \frac{(1+\mu\sin\theta)^2}{\sin\theta}\frac{dN}{d\theta} \tag{4-9}$$

Furthermore, as indicated by Novozhilov (1951) that the right side of (4-8) also may become infinity for some kinds of loadings, for example for the uniform pressure. To improve this situation a new function is defined by Novozhilov (1951) as follows:

$$Y = Z + i\kappa A\cot\theta \tag{4-10}$$

where A is a constant associated with the pressure.

The final equation expressed in terms of Y is

$$(1+\mu\sin\theta)\frac{d^2Y}{d\theta^2} - \mu\cos\theta\frac{dY}{d\theta} + i\kappa\sin\theta Y = \kappa\tilde{D}\cos\theta \tag{4-11}$$

where the complex function

$$\tilde{D} = -\left(\kappa A + i\mu\frac{pr_0}{2}\right) \tag{4-12}$$

It is worthy to indicate that the inhomogeneous term of (4-11) has no singularity at the transition points of $\theta = 0$ and $\theta = \pm\pi$. The advantage of Novozhilov's equation makes it easy to be used to find asymptotic solutions.

All of the force and moment resultants and the deformation components can be expressed in terms of Y. For simplification, these expressions are neglected.

The above derivation is given by Novozhilov (1951). Interested readers can refer to the Novozhilov's monograph in 1951 for more details.

4.3. A Novel Solution for Axisymmetric Toroidal Shells

In the section a set of new solutions is developed, which satisfies the accuracy of the thin shell theory and is numerically satisfactory. Furthermore, all of these new solutions are expressed in terms of generalized Airy functions. Namely, not only the homogeneous solutions but the particular solution is expressed in terms of generalized Airy functions. Moreover, three particular solutions are given, one of which is just the solution obtained by Tumarkin (1959) and Clark (1963).

Let us start from the Novozhilov's equation (4-11). Firstly, transform it to the equation

$$\frac{d^2y}{d\theta^2} + \left[\lambda^2 p_1(\theta) + p_2(\theta)\right]y = \lambda^2 G(\theta) \tag{4-13}$$

where

$$y = \frac{Y}{\sqrt{1+\mu\sin\theta}},$$

$$p_1(\theta) = \frac{\sin\theta}{1+\mu\sin\theta},$$

$$p_2(\theta) = -\frac{\mu\left[2\mu+2\sin\theta+\mu\cos^2\theta\right]}{4(1+\mu\sin\theta)^2},$$

$$G(\theta) = -\frac{i\tilde{D}\cos\theta}{(1+\mu\sin\theta)^{3/2}} \tag{4-14}$$

and

$$\lambda^2 = i\kappa \tag{4-15}$$

Note that λ is a complex large parameter.

4.3.1. Local Solutions

With Langer's transformation of

$$z = \varphi(\theta),\ \varphi(\theta) = \left[\frac{3}{2}\int_0^\theta \sqrt{p_1(\eta)}d\eta\right]^{2/3},\ \psi = \sqrt{\varphi'(\theta)y} \tag{4-16}$$

equation (4-13) becomes

$$\frac{d^2\psi}{dz^2} + \left[\lambda^2 z + \lambda R(z,\lambda)\right]\psi = \lambda^2 G(z,\lambda) \tag{4-17}$$

in which $R(z,\lambda)$ and $G(z,\lambda)$ can be expanded in the form:

$$R(z,\lambda) = \sum_{n=0}^{\infty}\lambda^{-n}R_n(z), G(z,\lambda) = \sum_{n=0}^{\infty}\lambda^{-n}G_n(z) \tag{4-18}$$

A comparison with equation (4-13) gives

$$R_1(z) = \frac{p_2}{\varphi'^2} + \frac{3}{4}\frac{\varphi''^2}{\varphi'^4} - \frac{1}{2}\frac{\varphi'''}{\varphi'^3},\ R_n = 0 \quad (n \neq 1) \tag{4-19$_1$}$$

$$G_0(z)=G(\theta),\ G_n=0 \quad (n\neq 0) \tag{4-19$_2$}$$

To determine the form of expansion of the homogenous solution of equation (4-17) or (4-13), we firstly find a "local solution", which is valid only in the vicinity of the transition point $z=0$. Therefore, we derive a "local equation"

Introduce

$$\zeta=-\lambda^{2/3}z \tag{4-20}$$

and define

$$r=R(0,\lambda)=\lambda^{-1}R_1(0),\ g=G(0,\lambda)=G_0(0)=-i\tilde{D} \tag{4-21}$$

Then, using constants r and g instead of $R(z,\lambda)$ and $G(z,\lambda)$, equation (4-17) is transferred to a "local equation" in the form:

$$\psi''-\zeta\psi+\lambda^{-1/3}r\psi=\lambda^{2/3}g \tag{4-22}$$

where prime indicates the differentiation with respect to ζ.

4.3.2. Comparison Equation and Generalized Airy Functions

Substituting the asymptotic series of

$$\psi(\zeta,\lambda)=\sum_{n=0}^{\infty}\left(\lambda^{-1/3}\right)^n\psi_n(\zeta) \tag{4-23}$$

into the homogeneous part of (4-22) and equating the coefficients of like powers of $\lambda^{-1/3}$, we obtain

$$\boldsymbol{A}\psi_0=0 \tag{4-24}$$

and

$$\boldsymbol{A}\psi_n=-r\psi_{n-1}\ (n\geq 1) \tag{4-25}$$

where the differential operator $\boldsymbol{A}$ is defined by Drazin and Reid (1981) as follows:

$$\boldsymbol{A}=\boldsymbol{D}^2-\zeta \text{ and } \boldsymbol{D}=\frac{d}{d\zeta} \tag{4-26}$$

The second order equation (4-24) is referred to as a comparison equation. Obviously, the comparison equation is an Airy equation. However, we prefer to use the generalized Airy functions $A_k(\zeta,p)$, $B_k(\zeta,p)$ and $B_0(\zeta,p)$ as the solutions of the Airy equation in stead of the ordinary Airy functions $Ai(\zeta)$ and $Bi(\zeta)$. The generalized Airy functions $A_k(\zeta,p)$, $B_k(\zeta,p)$ and $B_0(\zeta,p)$ have been given in (A2), (A3) and (A4).

It is easy to justify from (A11) that $A_k(\zeta,p)$ satisfies the equation

$$\boldsymbol{A}A_k(\zeta,p-1) = -(p-1)A_k(\zeta,p) \tag{4-27}$$

and $B_0(\zeta,p)$ and $B_k(\zeta,p)$ also satisfy (4-27) with $A_k(\zeta,p)$ replaced by $B_0(\zeta,p)$ and $B_k(\zeta,p)$. The asymptotic expansions of $B_k(\zeta,p)$ in T_k (see A15) are

$$B_k(\zeta,p) \sim (-1)^{1-p}(-p)!\zeta^{p-1}\left\{1-\frac{1}{3}(p-1)(p-2)(p-3)\zeta^3+\ldots\right\} \tag{4-28}$$

4.3.3. Expansion of Homogeneous Solution

It is seen from (4-27) by putting $p=1$ that the solution of equation (4-24) is

$$\psi_0 = A_K(\zeta,0) \tag{4-29}$$

where $k=1,2,3$, the same below. Substituting (4-29) into (4-25) with $n=1$ and comparing with (4-27) with $p=0$, we obtain

$$\psi_1 = -rA_k(\zeta,-1) \tag{4-30}$$

Similarly, substituting (4-30) into (4-25) with $n=2$ and comparing with (4-27) with $p=-1$ yields

$$\psi_2 = \frac{1}{2}(-r)^2 A_k(\zeta,-2) \tag{4-31}$$

Repetition of this process then gives

$$\psi_n = \frac{1}{n!}(-r)^n A_k(\zeta,-n),\ (n\geq 0) \tag{4-32}$$

Thus, substituting (4-32) into (4-23) gives

$$\psi(\zeta,\lambda)=\sum_{n=0}^{\infty}\left(\lambda^{-1/3}\right)^{n}\frac{1}{n!}(-r)^{n}A_{k}(\zeta,-n) \tag{4-33}$$

However, it is can be seen from the recursion formula (A11) that all of $A_k(\zeta,p)$, $(p\leq 0)$ can be expressed as a linear combination of $A_k(\zeta,0)$ and $A_k(\zeta,\pm 1)$. Then (4-33) is rewritten as

$$\psi^{(k)}(\zeta,\lambda)=\alpha(0,\lambda)A_k(\zeta,0)+\lambda^{-1/3}\beta(0,\lambda)A_k(\zeta,-1)\ (k=2,3) \tag{4-34}$$

where $\alpha(0,\lambda)$ and $\beta(0,\lambda)$ indicate that they are made up of the coefficients in equation (4-17) which are evaluated at the transition point $z=0$. They are given constants and possess expansions for λ^{-1} in the form

$$\alpha(0,\lambda)=\sum_{n=0}^{\infty}\alpha_n(0)\lambda^{-n}\ ,\ \beta(0,\lambda)=\sum_{n=0}^{\infty}\beta_n(0)\lambda^{-n} \tag{4-35}$$

In addition, the superscript of $\psi^{(k)}$ in (4-34) emphasizes that $\psi^{(k)}$ are actually three solutions. Of course, we can obtain two independent solutions from them in three possible manners. It will be shown later that the choice of $\psi^{(2)}$ and $\psi^{(3)}$ is more attractive.

The local expression (4-34) may help us to recognize that the uniformly valid solution of equation (4-17) has the expansion

$$\psi^{(k)}(\zeta,\lambda)=\alpha(z,\lambda)A_k(\zeta,0)+\lambda^{-1/3}\beta(z,\lambda)A_k(\zeta,-1) \tag{4-36}$$

where $\alpha(z,\lambda)$ and $\beta(z,\lambda)$ are the coefficients to be determined. Similar to (4-34), they are written in the form of the asymptotic series as

$$\alpha(z,\lambda)=\sum_{n=0}^{\infty}\alpha_n(z)\lambda^{-n}\ ,\ \beta(z,\lambda)=\sum_{n=0}^{\infty}\beta_n(z)\lambda^{-n} \tag{4-37}$$

Expansion (4-36) is the same as assumed by Olver (see Nayfeh (1973), equation (7.3.85)).

4.3.4. Expansion of Particular Solution

Similarly we will find the "local particular solution", which is valid only in the vicinity of the transition point $z=0$. According to the method of variation parameter, the local particular solution is

$$\psi^{*}(\zeta,\lambda)=c_1(\zeta)\psi^{(k)}(\zeta,\lambda)+c_2(\zeta)\psi^{(k+1)}(\zeta,\lambda) \tag{4-38}$$

where the superscript k is enumerated modulo 3. $\psi^{(k)}$ and $\psi^{(k+1)}$ are any linearly independent set of homogeneous solutions of equation (4-22) as given in (4-34).

Substituting (4-38) into equation (4-22) gives

$$\frac{dc_1(\zeta)}{d\zeta}=-\frac{\psi^{(k+1)}}{\mathcal{W}(\psi^{(k)},\psi^{(k+1)})}\lambda^{2/3}g \qquad (4\text{-}39)_1$$

and

$$\frac{dc_2(\zeta)}{d\zeta}=\frac{\psi^{(k)}}{\mathcal{W}(\psi^{(k)},\psi^{(k+1)})}\lambda^{2/3}g \qquad (4\text{-}39)_2$$

where $\mathcal{W}(\psi^{(k)},\psi^{(k+1)})$ denotes the Wronskians of $\psi^{(k)}$ and $\psi^{(k+1)}$. Substituting (4-34) into this Wronskians and noting $(A5)_1$, (A11), (A17) and (A10), we have

$$\mathcal{W}(\psi^{(k)},\psi^{(k+1)})=-\frac{[\alpha(0,\lambda)]^2}{2\pi i} \qquad (4\text{-}40)$$

Substituting (4-40) into (4-39) and integrating gives

$$c_1(\zeta)=\frac{2\pi i}{[\alpha(0,\lambda)]^2}\lambda^{2/3}g\left[\alpha(0,\lambda)A_{k+1}(\zeta,1)+\lambda^{-1/3}\beta(0,\lambda)A_{k+1}(\zeta,0)\right] \qquad (4\text{-}41)_1$$

and

$$c_2(\zeta)=\frac{2\pi i}{[\alpha(0,\lambda)]^2}\lambda^{2/3}g\left[\alpha(0,\lambda)A_k(\zeta,1)+\lambda^{-1/3}\beta(0,\lambda)A_k(\zeta,0)\right] \qquad (4\text{-}41)_2$$

Substituting (4-41) into (4-38) and noting $(A5)_1$, (A11), (A17) and (A10) then gives

$$\psi^*(\zeta,\lambda)=g\left[\frac{\beta(0,\lambda)}{\alpha(0,\lambda)}\right]^2+\lambda^{2/3}gB_{k+2}(\zeta,0)+\lambda^{1/3}g\frac{\beta(0,\lambda)}{\alpha(0,\lambda)}B_{k+2}(\zeta,-1) \qquad (4\text{-}42)$$

which is a particular solution of (4-22) and is valid only in the vicinity of the transition point $z=0$. Expression (4-42) guides us to assume the uniformly valid solution of equation (4-17) to be

$$\psi^*(\zeta,\lambda)=c(z,\lambda)+\lambda^{2/3}a(z,\lambda)B_{k+2}(\zeta,0)+\lambda^{1/3}b(z,\lambda)B_{k+2}(\zeta,-1) \qquad (4\text{-}43)$$

where $a(z,\lambda)$, $b(z,\lambda)$ and $c(z,\lambda)$ are coefficients to be determined and have expansions

$$a(z,\lambda)=\sum_{n=0}^{\infty}a_n(z)\lambda^{-n},\ b(z,\lambda)=\sum_{n=0}^{\infty}b_n(z)\lambda^{-n},\ c(z,\lambda)=\sum_{n=0}^{\infty}c_n(z)\lambda^{-n} \tag{4-44}$$

It can be seen from (4-43) that we have actually obtained three particular solutions of equation (4-17), since the subscript k may be equal to 1, 2 and 3, respectively.

Each of them corresponds to one of three linearly independent sets of solutions denoted by (4-36).

It can seen from (A3) and Figure A.1 that only B_1 is a real integration among B_1, B_2 and B_3. Thus, putting $k=2$ in (4-43) gives a simplest particular solution

$$\psi^*(\zeta,\lambda)=c(z,\lambda)+\lambda^{2/3}a(z,\lambda)B_1(\zeta,0)+\lambda^{1/3}b(z,\lambda)B_1(\zeta,-1) \tag{4-45}$$

Correspondingly, putting $k=2$ in (4-38) gives a linearly independent set of solutions $\psi^{(2)}$ and $\psi^{(3)}$, which associated with the particular solution (4-43).

Hence, two independent homogeneous solutions of equation (4-17), see (4-36), have to be chosen as

$$\psi^{(k)}(\zeta,\lambda)=\alpha(z,\lambda)A_k(\zeta,0)+\lambda^{-1/3}\beta(z,\lambda)A_k(\zeta,-1)\ \ (k=2,3) \tag{4-46}$$

As mentioned previously, Tumarkin (1959) obtained a complete asymptotic expansion of equation (4-17):

$$v=C(z,\lambda)+\lambda^{2/3}A(z,\lambda)T(\xi)+\lambda^{1/3}B(z,\lambda)T'(\xi),\ \ \xi=\lambda^{2/3}z \tag{4-47}$$

where v is transformed from y, $T(\xi)$ is defined as the solution of equation

$$T''+\xi T=1,\ T=\frac{1}{\xi}\ \ as\ |\xi|\to\infty \tag{4-48}$$

Thus, $T(\xi)$ may be represented in terms of Lommel function as

$$T(\xi)=\frac{2}{3}\xi^{1/2}S_{0,1/3}\left(\frac{2}{3}\xi^{3/2}\right)=\int_0^{\infty}\exp\left(-\xi t-\frac{1}{3}t^3\right)dt \tag{4-49}$$

It is remarkable that, a comparison of (A3) with (4-49) gives

$$B_1(\zeta,0)=\int_0^x\exp(\zeta t-t^3/3)dt=T(-\zeta) \tag{4-50}$$

and

$$B_1(\zeta,-1) = B_1'(\zeta,0) = -T'(-\zeta) \tag{4-51}$$

Therefore, our particular solution (4-45) is actually identical to the famous solution obtained by Tumarkin(1959) and Clark(1963).

4.4. Final results

The particular solution is expressed by (4-45) and two homogeneous solutions are (4-46).

4.4.1. Particular Solution

Within the accuracy extension of thin shell theory, the particular solution (4-45) is of the form

$$\begin{aligned}\psi^*(\zeta,\lambda) &= \left(c_0+\lambda^{-1}c_1\right)+\lambda^{2/3}\left(a_0+\lambda^{-1}a_1\right)B_1(\zeta,0)+\lambda^{1/3}\left(b_0+\lambda^{-1}b_1\right)B_1(\zeta,-1)\\ &= c_0+\lambda^{2/3}a_0B_1(\zeta,0)+\lambda^{1/3}b_0B_1(\zeta,-1)\\ &\quad+\lambda^{-1}\left[c_1+\lambda^{2/3}a_1B_1(\zeta,0)+\lambda^{1/3}b_1B_1(\zeta,-1)\right]\end{aligned} \tag{4-52}$$

in which

$$\psi_0^*(\zeta,\lambda) = c_0+\lambda^{2/3}a_0B_1(\zeta,0) \sim \mathrm{O}\left(h^0\right) \tag{4-53}$$

is the first approximation and

$$\begin{aligned}\psi_{high}^*(\zeta,\lambda) &= \lambda^{1/3}b_0B_1(\zeta,-1)+\lambda^{-1}\left[c_1+\lambda^{2/3}a_1B_1(\zeta,0)\right]\\ &= \lambda^{-1}\left[c_1+\lambda^{2/3}a_1B_1(\zeta,0)+\lambda^{4/3}b_0B_1(\zeta,-1)\right] \sim \mathrm{O}\left(h^{1/2}\right)\end{aligned} \tag{4-54}$$

is the high order approximation.

Substituting (4-52) into equation (4-17), equating coefficients of $B_1(\zeta,0)$ and $B_1(\zeta,-1)$ to zero, respectively, successive equations about a_n , b_n and c_n $(n=0,1)$ can be obtained in the form:

$$-2b_0'+a_1+zc_1=0 \tag{4-55}$$

$$a_0+zc_0-G_0=0 \tag{4-56}$$

$$2zb_0'+b_0=0 \tag{4-57}$$

$$b_0''+b_0R_1-2a_1'=0 \tag{4-58}$$

$$a_0' = 0 \tag{4-59}$$

Equation (4-59) gives

$$a_0 = const$$

Then, according to (4-56), to ensure c_0 a finite value at $z=0$, it is necessary that

$$a_0 = G_0(0) \tag{4-60}$$

A substitution of a_0 back into (4-56) yields

$$c_0 = \frac{G_0(z) - G_0(0)}{z} \tag{4-61}$$

The solution of (4-57) is

$$b_0 = \frac{C}{\sqrt{z}} \tag{4-62}$$

where C is a constant to be determined. Obviously, C must be zero, otherwise $b_0(z)$ becomes infinite at $z=0$, namely

$$b_0 = 0 \tag{4-63}$$

It can be seen from (4-55) and (4-63) that c_1 is finite at $z=0$ as long as

$$a_1(0) = 0$$

Then, substituting $a_1(0)=0$ into (4-59) gives

$$a_1 = 0 \tag{4-64}$$

Substituting (4-63) and (4-64) into (4-55) gives

$$c_1 = 0 \tag{4-65}$$

As a result, the first approximation to the particular solution of (4-17) or (4-13) is

$$\psi_0^*(\zeta,\lambda)=\frac{G_0(z)-G_0(0)}{z}+\lambda^{2/3}G_0(0)B_1(\zeta,0)\sim O(h^0) \tag{4-66}$$

and the high order approximation to the particular solution is zero. The complete particular solution is

$$\psi^*(\zeta,\lambda)=\psi_0^*=\frac{G_0(z)-G_0(0)}{z}+\lambda^{2/3}G_0(0)B_1(\zeta,0) \tag{4-67}$$

In other words, the first approximation is also a complete approximation to the particular solution in the axisymmetric case of the toroidal shells.

It is worthy to indicate that $\dfrac{G_0(z)-G_0(0)}{z}$ in the particular solution (4-67) is a membrane solution. Actually, letting $\lambda\to\infty$ in equation (4-17) gives the membrane equation: $z\psi^{(0)}=G(z,\lambda)$, whose solution is just $\dfrac{G_0(z)-G_0(0)}{z}$ under the analytic condition at the transition point $z=0$.

Expression (4-67) agrees with the result obtained by Holstein (1950), Clark (1958, 1963) and Tumarkin (1959) (see Nayfeh (1970), equation (7.3.125)).

4.4.2. Homogeneous Solutions

Within the accuracy extension of thin shell theory, the homogeneous solutions (4-46) are of the form

$$\psi^{(k)}(\zeta,\lambda)=\left(\alpha_0+\lambda^{-1}\alpha_1\right)A_k(\zeta,0)+\lambda^{-1/3}\left(\beta_0+\lambda^{-1}\beta_1\right)A_k(\zeta,-1) \tag{4-68}$$

$$(k=2,3)$$

in which

$$\psi_0^{(k)}(\zeta,\lambda)=\alpha_0A_k(\zeta,0)+\lambda^{-1/3}\beta_0A_k(\zeta,-1)\sim O\left(h^{1/12}\right) \tag{4-69}$$

is the first approximation and

$$\psi_{high}^{(k)}(\zeta,\lambda)=\lambda^{-1}\left[\alpha_1A_k(\zeta,0)+\lambda^{-1/3}\beta_1A_k(\zeta,-1)\right]\sim O\left(h^{7/12}\right) \tag{4-70}$$

is the high order approximation.

Substituting (4-68) into equation (4-17), equating coefficients of $A_k(\zeta,0)$ and $A_k(\zeta,-1)$ to zero, respectively, successive equations about α_n and β_n $(n=0,1)$ can be obtained in the form:

$$\alpha_0'' + R_1\alpha_0 + 2z\beta_1' + \beta_1 = 0 \tag{4-71}$$

$$2z\beta_0' + \beta_0 = 0 \tag{4-72}$$

$$\beta_0'' + R_1\beta_0 - 2\alpha_1' = 0 \tag{4-73}$$

$$\alpha_0' = 0 \tag{4-74}$$

Integrating (4-74) gives $\alpha_0 = const$. Without losing generality, we take

$$\alpha_0 = 1 \tag{4-75}$$

The solution of equation (4-72) is $\beta_0(z) = \dfrac{\beta_0(0)}{\sqrt{z}}$. To avoid an infinite value at $z = 0$, it is necessary that $\beta_0(0) = 0$, which indicates

$$\beta_0(z) = 0 \tag{4-76}$$

Substituting (4-75) into (4-71) gives $R_1\alpha_0 + 2z\beta_1' + \beta_1 = 0$, whose solution is

$$\beta_1(z) = -\frac{\alpha_0}{2\sqrt{z}}\int_0^z \frac{R_1(\tau)}{\sqrt{\tau}}d\tau \tag{4-77}$$

Substituting (4-76) into (4-73) gives $\alpha_1 = const$. Without losing generality, we take

$$\alpha_1 = 0 \tag{4-78}$$

By substituting (4-75) and (4-76) into (4-69), the first approximation to the homogeneous solution is obtained

$$\psi_0^{(k)}(\zeta,\lambda) = A_k(\zeta,0) \tag{4-79}$$

By substituting (4-77) and (4-78) into (4-70), the high order approximation to the homogeneous solution is obtained

$$\psi_{high}^{(k)}(\zeta,\lambda) = \lambda^{-4/3}\beta_1 A_k(\zeta,-1) \tag{4-80}$$

The complete homogeneous solution is

$$\psi^{(k)}(\zeta,\lambda)=A_k(\zeta,0)-\lambda^{-4/3}\left[\frac{1}{2\sqrt{z}}\int_0^z\frac{R_1(\tau)}{\sqrt{\tau}}d\tau\right]A_k(\zeta,-1),\ (k=2,3) \tag{4-81}$$

Finally, the complete solution is

$$\begin{aligned}\psi&=C_k\psi^{(k)}+\psi^*\\&=\left[C_2A_2(\zeta,0)+C_3A_3(\zeta,0)\right]\\&\quad-\lambda^{-4/3}\left[\frac{1}{2\sqrt{z}}\int_0^z\frac{R_1(\tau)}{\sqrt{\tau}}d\tau\right]\left[C_2A_2(\zeta,-1)+C_3A_3(\zeta,-1)\right]\\&\quad+\frac{G_0(z)-G_0(0)}{z}+\lambda^{2/3}G_0(0)B_1(\zeta,0)\end{aligned} \tag{4-82}$$

in which two complex constants C_2 and C_3 can be used to satisfy 4 boundary conditions.

Novozhilov (1951) gave and only gave a first approximation to the homogeneous solution in terms of Bessel's functions of order $\frac{1}{3}$: $h^{(1)}_{\frac{1}{3}}$ and $h^{(2)}_{\frac{1}{3}}$ instead of $A_2(\zeta,0)$ and $A_3(\zeta,0)$. (see Novozhilov (1951), equation(4.11.26)) However, $A_2(\zeta,0)$ and $A_3(\zeta,0)$ and the first-class of Airy functions Ai are related by

$$A_2(\zeta,0)=e^{2\pi i/3}Ai\left(\zeta e^{2\pi i/3}\right) \tag*{(4-83)$_1$}$$

$$A_3(\zeta,0)=e^{-2\pi i/3}Ai\left(\zeta e^{-2\pi i/3}\right) \tag*{(4-83)$_2$}$$

The first-class of Airy functions can be expressed in terms of Bessel's functions of order $\frac{1}{3}$. Therefore, our first approximation to the homogeneous solution can be considered same as the Novozhilov' results.

5. Axisymmetric Deformation of Orthotropic Toroidal Shells

There are various equations governing the axisymmetric deformation of orthotropic toroidal shells. Wang et al. (1986) gave an equation in terms of a complex rotation Θ, whose real part is a rotation of a meridian element, as follows:

$$\frac{d^2\Theta}{d\theta^2}+p_0\frac{d\Theta}{d\theta}+\left[\left(i\kappa^+-\mu\right)p_1+p_2\right]\Theta=i\kappa^+p_3 \tag{5-1}$$

where

$$p_0(\theta)=\frac{\mu\cos\theta}{1+\mu\sin\theta},\ p_1(\theta)=\frac{\sin\theta}{1+\mu\sin\theta},\ p_2(\theta)=-\left[p_0(\theta)\right]^2$$

$$p_3=R_0\left(\frac{Q_1^*}{K}-\frac{1}{E_2hr_0}\frac{dT^*}{d\theta}\right),\ \kappa^+=\sqrt{12\left(1-\nu_1\nu_2\right)\frac{E_2}{E_1}\frac{\mu r_0}{h}}$$

$$K=ih^2\sqrt{\frac{E_1E_2}{12\left(1-\nu_1\nu_2\right)}},\ \mu=\frac{r_0}{R_0}$$

$$T^*=\frac{\sin\theta}{R_0\sigma}q_n-\frac{R_0}{\sigma}\frac{1}{\sin^2\theta}\int_0^\theta\left(q_n\cos\theta-q_1\sin\theta\right)\sigma d\theta-\frac{\sin\theta}{R_0\sigma}B$$

$$Q_1^*=R_0B\frac{\sigma\cos\theta}{\sin^2\theta} \tag{5-2}$$

and E_1, E_2 and ν_1, ν_2 are Yang's module and Poison's ratio along two principal curvature, respectively, and B is a real constant.

When Θ is solved out, all the force and moment resultants and the deformation components can be expressed in terms of Θ.

Note that $p_1=0$ when $\theta=0$ and $\theta=\pm\pi$, it implies a presence of the transition points at $\theta=0$ and $\theta=\pm\pi$.

In fact, with Langer's transformation of

$$\Theta=\phi U,\ z(\theta)=\left[\frac{3}{2}\int_0^\theta\sqrt{p_1(\eta)}d\eta\right]^{2/3},\ \phi=\left(\frac{z}{\sin\theta\left(1+\mu\sin\theta\right)}\right)^{1/4} \tag{5-3}$$

equation (5-1) becomes

$$\frac{d^2U}{dz^2}+\left[\lambda^2z+R_1(z)\right]U=\lambda^2G_0(z) \tag{5-4}$$

where

$$\lambda^2=i\kappa^+ \tag{5-5}$$

$$R_1(z)=\frac{z}{\phi\sin\theta}\frac{d}{d\theta}\left[(1+\mu\sin\theta)\frac{d\phi}{d\theta}\right]-\left(\frac{\mu\cos\theta}{1+\mu\sin\theta}\right)^2\phi \tag{5-6}$$

$$G_0(z)=R_0\frac{z(1+\mu\sin\theta)}{\phi\sin\theta}\left(\frac{Q_1^*}{K}-\frac{1}{E_2hr_0}\frac{dT^*}{d\theta}\right) \tag{5-7}$$

It is seen that equation (5-4) is similar to (4-17). Unfortunately, $R_1(z)$ in (5-4) was neglected by Wang et al. (1986). As the result, only a first approximation to the homogeneous solutions was given in their paper.

5.1. Homogeneous Solutions

In analogy with section 4, a uniformly valid homogeneous solution of equation (5-4) that satisfies the accuracy of thin shell theory is (see (4-81))

$$U^{(k)}(\zeta,\lambda)=\alpha_0A_k(\zeta,0)+\lambda^{-4/3}\beta_1(z)A_k(\zeta,-1)\ \ (k=2,3) \tag{5-8}$$

where

$$\zeta=-\lambda^{2/3}z \tag{5-9}$$

$$\alpha_0=1\,,\ \beta_1=-\frac{1}{2\sqrt{z}}\int_0^z\frac{R_1(\tau)}{\sqrt{\tau}}d\tau \tag{5-10}$$

Finally, a complete homogeneous solution of equation (5-1) that satisfies the accuracy of thin shell theory is

$$\Theta=\phi\sum_{k=2}^{3}C_kU^{(k)}=\phi\left\{\sum_{k=2}^{3}C_kA_k(\zeta,0)-\lambda^{-4/3}\left[\frac{1}{2\sqrt{z}}\int_0^z\frac{R_1(\tau)}{\sqrt{\tau}}d\tau\right]\sum_{k=2}^{3}C_kA_k(\zeta,-1)\right\} \tag{5-11}$$

Obviously, the first summation in the right side of the above equation is a first approximation to the homogeneous solutions, which coincides with the result given by Wang et al. (1986). The second summation is a higher approximation to the homogeneous solutions which was neglected by Wang et al. (1986).

5.2. Particular Solution

In the same way, in accordance with (4-67), the particular solution of equation (5-4) or (5-1) is

$$\Theta = \phi\psi^*(\zeta,\lambda) = \phi\left\{\frac{G_0(z)-G_0(0)}{z} + \lambda^{2/3}G_0(0)B_1(\zeta,0)\right\} \quad (5\text{-}12)$$

which is identical to what Wang et al. (1986) obtained as long as the general Airy function $B_1(\zeta,0)$ is replaced by Lommel functions.

Obviously, if $E_1 = E_2 = E$ and $\nu_1 = \nu_2 = \nu$, then κ^+ equals κ in section 4. As the result, all the equations and their solutions for orthotropic toroidal shells in this section become corresponding ones in section 4.

CONCLUSION

1) All the asymptotic solutions (homogeneous and particular solutions) for a toroidal shell subjected to arbitrary loadings (axisymmetric and non-axisymmetric loadings) are given in the present paper, which are uniformly valid everywhere in the entire shell including the transition point and satisfy the accuracy of the thin shell theory.
2) In the non-axisymmetric analysis, the large parameter λ becomes larger as m increases based on the formula $\lambda^2 = i\sqrt{12(1-\nu^2)}\frac{\mu r_0}{h}m^4$.
   This ensures a more accurate analysis for large m. However, as mentioned in section 3, the results are also exact for the smallest value of m ($m=1$). So that it can be concluded that our solutions for the non-axisymmetric loadings are sufficiently exact for all the values of m.
3) The axisymmetric solutions are given in (4-82), in which the high-order approximation to the homogeneous solutions is presented for the first time.
4) In the non-axisymmetric case, the particular solution $\psi^*(z)$ satisfies the membrane equation (3-49). This fact shows that non-axisymmetric surface loadings (normal components) are balanced totally by the membrane stresses.
   This differs completely from the axisymmetric case in which the axisymmetric surface loadings are balanced by both the membrane stresses and the bending moments (see equation (4-67)).
5) In the non-axisymmetric case, there is a membrane solution as a homogeneous solution, which is a slowly varying function (see (3-55)).
   This fact indicates that the boundary conditions are satisfied partially by membrane stresses. Contrarily, there is not any membrane solution as a homogeneous solution (see (4-81)) in the axisymmetric case. All the two homogeneous solutions are fast varying functions.
   This fact indicates that the boundary conditions are satisfied totally by bending moments in the axisymmetric case.

APPENDIX

According to Drazin and Reid (1981), we find that the differential equation

$$\left(\boldsymbol{D}^3 - \zeta \boldsymbol{D} + p - 1\right)u = 0 \tag{A1}$$

has three solutions as follows:

$$A_k(\zeta, p) = \frac{1}{2\pi i}\int_{L_k} t^{-p}\exp\left(\zeta t - \frac{1}{3}t^3\right)dt \quad (k = 1,2,3; p = 0, \pm 1, \pm 2, ...) \tag{A2}$$

$$B_k(\zeta, p) = \frac{1}{2\pi i}\int_{I_k} t^{-p}\exp\left(\zeta t - \frac{1}{3}t^3\right)dt \quad (k = 1,2,3; p = 0, -1, -2, ...) \tag{A3}$$

$$B_0(\zeta, p) = \frac{1}{2\pi i}\int_{L_0} t^{-p}\exp\left(\zeta t - \frac{1}{3}t^3\right)dt \quad (p = 0, \pm 1, \pm 2, ...) \tag{A4}$$

in which the integral paths are shown in Figure A.1

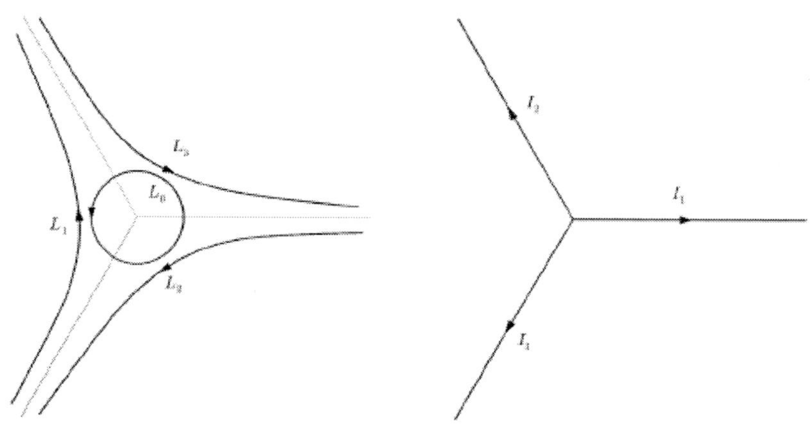

Figure A.1. Paths of integration in the t-plane.

The derivatives of these solutions satisfy the relations

$$\boldsymbol{D}^n A_k(\zeta, p) = A_k(\zeta, p - n) \tag{A5$_1$}$$

$$\boldsymbol{D}^n B_k(\zeta, p) = B_k(\zeta, p - n) \tag{A5$_2$}$$

$$\boldsymbol{D}^n B_0(\zeta, p) = B_0(\zeta, p - n) \tag{A5$_3$}$$

Differentiating A1 and noting that

$$\boldsymbol{D}\left(\boldsymbol{D}^3-\zeta\boldsymbol{D}+p-1\right)=\boldsymbol{TD}+(p+1)\boldsymbol{D} \tag{A6}$$

we have

$$\boldsymbol{TD}A_k(\zeta,p+1)=-(p+2)A_k(\zeta,p) \tag{A7}$$

$$\boldsymbol{TD}B_k(\zeta,p+1)=-(p+2)B_k(\zeta,p) \tag{A8}$$

$$\boldsymbol{TD}B_0(\zeta,p+1)=-(p+2)B_0(\zeta,p) \tag{A9}$$

Note from equation (A3) that $B_0(\zeta,p)\equiv 0$, if $p\le 0$. Otherwise, it is a polynomial in ζ of degree $p-1$, which, by the residue theorem, is simply the coefficient of t^{p-1} in the expansion of $\exp\left(\zeta t-\frac{1}{3}t^3\right)$. The first few terms of these polynomials are

$$\begin{aligned}
&B_0(\zeta,1)=1,\ B_0(\zeta,4)=\frac{1}{3!}\zeta^3-\frac{1}{3}\\
&B_0(\zeta,2)=\zeta,\ B_0(\zeta,5)=\frac{1}{4!}\zeta^4-\frac{1}{3}\zeta\\
&B_0(\zeta,3)=\frac{1}{2!}\zeta^2,\ B_0(\zeta,6)=\frac{1}{5!}\zeta^5-\frac{1}{6}\zeta^2
\end{aligned} \tag{A10}$$

By using (A7) we immediately obtain the particular integrals of the inhomogeneous equation $\boldsymbol{TD}u=A_k(\zeta,p)$ for all values of p except $p=-2$, and have the recursion formula

$$A_k(\zeta,p-3)-\zeta A_k(\zeta,p-1)+(p-1)A_k(\zeta,p)=0 \tag{A11}$$

Thus, for other values of p, $A_k(\zeta,p)$ can be expressed as a linear combination of $A_k(\zeta,0)$ and $A_k(\zeta,\pm 1)$ with polynomial coefficients.

The asymptotic behavior of $A_k(\zeta,p)$ is given in the form

$$A_1(\zeta,p)\sim A_-(\zeta,p)\quad (\zeta\in T_2\cup T_3)$$

$$A_2(\zeta,p)\sim iA_+(\zeta,p)\ \ (\zeta\in T_3\cup T_1) \tag{A12}$$

$$A_3(\zeta,p)\sim\begin{cases}A_-(\zeta,p) & (\zeta\in T_1)\\ iA_+(\zeta,p) & (\zeta\in T_2)\end{cases}$$

and

$$\begin{bmatrix}A_1(\zeta,p)\\A_2(\zeta,p)\\A_3(\zeta,p)\end{bmatrix}\sim\begin{bmatrix}-i & -1 & -1\\ i & -1 & -1\\ -i & -1 & -1\end{bmatrix}\begin{bmatrix}A_+(\zeta,p)\\B_0(\zeta,p)\\A_-(\zeta,p)\end{bmatrix}\ \ (\zeta\in T_k, p\in Z) \tag{A13}$$

where T_k are the sectors shown in Figure A.2 and

$$A_\pm(\zeta,p)=\frac{1}{2\sqrt{\pi}}(\pm1)^p\,\zeta^{-(2p+1)/4}\exp\left(\pm\frac{2}{3}\zeta^{3/2}\right)\sum_{s=0}^{\infty}(\pm1)^s a_s(p)\zeta^{-s} \tag{A14}$$

in which

$$a_0(p)=1$$
$$a_1(p)=\frac{1}{2^3 3^2}\left(12p^2+24p+5\right)$$

$$a_2(p)=\frac{1}{2^7 3^4}\left(144p^4+1344p^3+3864p^2+3504p+3855\right)+\ldots$$

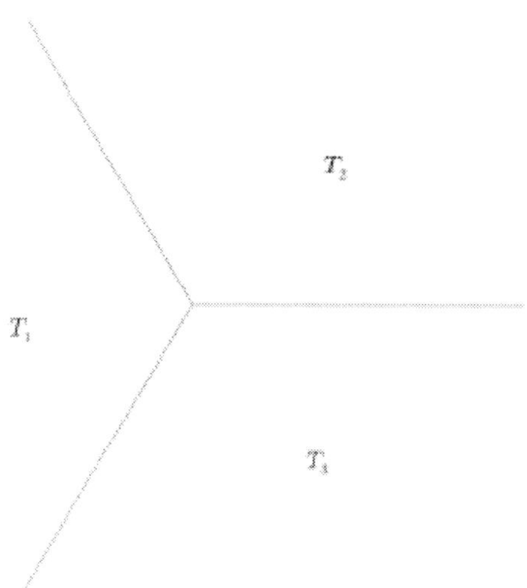

Figure A.2. Sectors for the Airy functions in the ζ −plane.

The asymptotic expansions of $B_k(\zeta, p)$ in T_k (see Figure A.2) are

$$B_k(\zeta,p) \sim (-1)^{1-p}(-p)!\zeta^{p-1}\left\{1-\frac{1}{3}(p-1)(p-2)(p-3)\zeta^3+\ldots\right\} \tag{A15}$$

The Wronskian of $A_k(\zeta, p)$ is given by Drazin and Reid (1981) as follows:

$$\mathcal{W}(A_k, A_{k+1}) = \begin{cases} -(-p)!(-1)^p \dfrac{1}{2\pi i} B_0(\zeta, 1-p) & (p = 0,-1,-2,\ldots) \\ -\dfrac{1}{(p-1)!}\dfrac{1}{2\pi i} B_{k+2}(\zeta,1-p) & (p = 1,2,3,\ldots) \end{cases} \tag{A16}$$

where the subscript k is enumerated modulo 3.

Functions $B_k(\zeta, p)$ also satisfy (A11) with $A_k(\zeta, p)$ replaced by $B_k(\zeta, p)$ such that

$$B_k(\zeta,p-3)-\zeta B_k(\zeta,p-1)+(p-1)B_k(\zeta,p)=0 \quad (p\le 0) \tag{A17}$$

The other important recursion formula is

$$B_k(\zeta,-2)-\zeta B_k(\zeta,0)-1=0 \tag{A18}$$

In order to obtain the particular integrals of the non-homogeneous equation $\boldsymbol{TD}u = A_k(\zeta,-2)$, we consider the following generalized Airy functions defined by Drazin and Reid (1981):

$$A_k(\zeta,p,q)=\frac{1}{2\pi i}\int_{L_k} t^{-p}(\ln t)^q \exp\left(\zeta t-\frac{1}{3}t^3\right)dt \tag{A19}$$

where $k = 1,2,3;\ p = 0,\pm1,\pm2,\ldots;\ q = 0,1,2,\ldots;$ and a branch cut has placed along the positive real axis in the $t-$plane so that $0 \le ph(t) \le 2\pi$. Similarly, the derivatives satisfy the relation

$$\boldsymbol{D}^n A_k(\zeta,p,q)=A_k(\zeta,p-n,q) \tag{A20}$$

Functions $A_k(\zeta,p,q)$ are the solutions of the non-homogeneous equation

$$\boldsymbol{TD}A_k(\zeta,p+1,q)=-(p+2)A_k(\zeta,p,q)+qA_k(\zeta,p,q-1) \tag{A21}$$

There is a similar recursion formula

$$A_k(\zeta, p-3, q) - \zeta A_k(\zeta, p-1, q) + (p-1)A_k(\zeta, p, q) = qA_k(\zeta, p, q-1) \tag{A22}$$

It is easily seen that

$$A_k(\zeta, p, 0) \equiv A_k(\zeta, p) \tag{A23}$$

The asymptotic expansion of $A_k(\zeta, p, 1)$ is

$$A_1(\zeta, p, 1) \sim \frac{1}{2\sqrt{\pi}}(-1)^p \zeta^{-(2p+1)/4} \exp\left(\frac{2}{3}\zeta^{3/2}\right) \quad (\zeta \in T_2 \cup T_3)$$
$$\times \sum_{s=0}^{\infty}(-1)^s \left[\left(\frac{1}{2}\ln\zeta + \pi i\right)a_s(p) - \frac{da_s}{dp}\right]\left(\frac{2}{3}\zeta^{3/2}\right)^s \tag{A24}$$

The corresponding expansions for $A_2(\zeta, p, 1)$ and $A_3(\zeta, p, 1)$ can be obtained from the recursion formula

$$A_2(\zeta, p, 1) = \mathrm{e}^{-2(p-1)\pi/3}\left[A_1\left(\zeta e^{2\pi i/3}, p, 1\right) + \frac{2}{3}\pi i A_1\left(\zeta e^{2\pi i/3}, p\right)\right] \tag{A25}$$

$$A_3(\zeta, p, 1) = \mathrm{e}^{2(p-1)\pi/3}\left[A_1\left(\zeta e^{-2\pi i/3}, p, 1\right) - \frac{2}{3}\pi i A_1\left(\zeta e^{-2\pi i/3}, p\right)\right] \tag{A26}$$

In order to deal with the third homogeneous the following generalized Airy functions defined by Drazin and Reid (1981) have to be used

$$B_k(\zeta, p, q) = \frac{1}{2\pi i}\int_{\infty \exp[2(k-1)\pi i/3]}^{(0_\pm)} t^{-p}(\ln t)^q \exp\left(\zeta t - \frac{1}{3}t^3\right)dt \tag{A27}$$

where $k = 1, 2, 3;\ p = 0, \pm1, \pm2, \ldots;\ q = 0, 1, 2, \ldots$. Note that

$$B_k(\zeta, p, 0) \equiv B_0(\zeta, p) \ \ (p = 0, \pm1, \pm2, \ldots) \tag{A28}$$

and

$$B_k(\zeta, p, 1) \equiv B_k(\zeta, p) \ \ (k = 1, 2, 3; p = 0, -1, -2, \ldots) \tag{A29}$$

Functions $B_k(\zeta, p, q)$ also satisfy (A20), (A21) and (A22) with $A_k(\zeta, p, q)$ replaced by $B_k(\zeta, p, q)$. The asymptotic expansion of $B_1(\zeta, 1, 1)$ in the sector T_1 is

$$B_1(\zeta, 1, 1) \sim -\ln\zeta - \gamma + \sum_{n=1}^{\infty} \frac{(3n-1)!}{3^n n!} \zeta^{-3n} \tag{A30}$$

where γ is the Euler constant.

About Author

R.J Zhang received the Ph.D. degree in Engineering Mechanics from Tsinghua University, Beijing, China, 1988. He is Professor of the School of Aerospace Engineering and Applied Mechanics, Tongji University, Shanghai, China. His research interests include the solid mechanics, structures with composite materials, theory of thin shells, interaction between structures and fluids, as well as the flow in electric fields.

Acknowledgments

Supporting by the China National Science Foundation under grant 11172214 is gratefully acknowledged. The present paper is consecrated in memory of Professor Zhang Wei, who is my supervisor for Ph.D.

References

Drazin, P.D. and Reid, W.H., Hydrodynamic stability, Cambridge University Press, Cambridge, 1981.

Nayfeh, A. H. , Perturbation methods, John Wiley and Sons. Inc., 1973.

Novozhilov,V.V., Theory of elastic shell, *State Press of Naval Architecture*, 1951. (Chinese translation).

Sanders, Jr. J. L., An improved first-approximation theory for thin shells, *NASA TR* R-24, 1959.

Sanders, Jr. J. L., On the shell equations in complex form, *NASA-CR*-89179, SM-16, 1967.

Steele C.R., Toroidal shells with nonsymmetric loading. A Dissertation for the Degree of Doctor of Philosophy of Stanford University, 1959.

Wang Anwen, Xia Zhixi, Ren Wenmin, The asymptotic analysis for the stresses in orthotropic toroidal shells under axially symmetric loads, *Journal of Tsinghua University*, Vol.26, No. 3, 65-79, 1986.

Xia Z. H. and Zhang W., The general solution for thin-walled curved tubes with arbitrary loadings and various boundary conditions, *Int. J. Press. Ves. Piping* 26, 129-144, 1986.

Zhang Wei, Dissertation T .H, Berlin, published partly in Science Report of Tsinghua University (in Cheness), Ser.A, Vol.5, 289-349, 1949.

Zhang, R.J. and Zhang W., Asymptotic solutions of toroidal shells under nonsymmetic Loading, *Science in China (Series A),* Vol.25, No. 6, 614-619, 1995.

Zhang, R.J. and Zhang W., Turning point solution for thin shell vibrations, *Int. J. Solids Structures*, 27 (10), 1311-1326, 1991.

Zhang, R.J., A novel solution of toroidal shells under axisymmetic Loading, *Applied Mathematics and Mechanics*, Vol.20, No.5, 519-526, 1999.

Zhang, R.J., Homogeneous solutions of orthotropic toroidal shells under axisymmetic Loading, *Chinese Journal of Applied Mechanics*, Vol.9, No.2, 84-89, 1992.

Zhang, R.J., Toroidal Shells under Nonsymmetic Loading, *Int.J. Solids structures,* Vol.31, No.19, 2735-2750, 1994.

In: Toroidal Shells
Editor: Bohua Sun
ISBN: 978-1-61942-247-6

Chapter 3

FINITE ELEMENT ANALYSIS FOR VIBRATION OF AXISYMMETRIC TOROIDAL SHELLS OF REVOLUTION

George R. Buchanan*
Department of Civil and Environmental Engineering,
Tennessee Technological University, Cookeville TN, US

NOMENCLATURE

The following symbols are used in this Chapter:

a=radius of thin toroidal shell measured to the mid-plane of the shell;
a,b=major and minor dimensions of an ellipse defined in Figure 5.1;
a,L=dimensions of an torus of oblong shape with spherical caps defined in Figure 5.3;
a_o=outside radius of thin toroidal shell, see Table 3.1;
A_1,A_2= Lamé parameters for the shell;
b=inside radius of shell as defined in Section 4;
[B]=finite element B matrix;
C_{11},C_{33},C_{12},C_{13},C_{44},C_{66}=elasticity material constants;
[C]=matrix of material constants;
$[C_r]_{cobalt}$= nondimensional material constants for cobalt poled relative to r;
$[C_\varphi]_{cobalt}$= nondimensional material constants for cobalt poled relative to φ;
C_{kj}=material constants in matrix format;
$\bar{C}_{kj}$=nondimensional material constants in matrix format;
{d}=matrix of finite nodal point unknowns;
D_{11},D_{12},D_{33}=material constants for shell curvatures;
$[D_b]$=matrix of material constants for shell curvatures;

* email: gbuchanan@tntech.edu How do you want to handle this footnote with regards to the death of Dr. Buchanan?

$[D_m]$=matrix of material constants for membrane strains;
$[D_s]$=matrix of material constants for shear strains;
$[\bar{D}]$=material matrix (see equation (2.17)and(2.20));
$[\bar{D}_r]$=nondimensional material matrix poled relative to r;
$[\bar{D}_\varphi]$=nondimensional material matrix poled relative to φ;
dV=differential volume;
$\vec{e}_r, \vec{e}_\vartheta, \vec{e}_\varphi$=unit vectors in the toroidal coordinate system;
E=Young's Modulus;
G=shear modulus;
G_{13},G_{23}=material constants for shear strains;
h=shell thickness;
$h_r, h_\vartheta, h_\varphi$=scale factors;
i=imaginary number;
$\vec{\imath}, \vec{\jmath}, \vec{k}$=unit vectors in the Cartesian coordinate system;
k'=shear correction coefficient;
k_{11},k_{12},k_{33}=material constants for membrane strains;
[K]=finite element stiffness matrix;
[L]=finite element operator matrix;
M_{11},M_{22},M_{12}=shell bending moments;
{M}=matrix of shell bending moments;
[M]=finite element mass matrix;
n=circumferential wave number;
N_{11},N_{22},N_{12}=shell membrane forces;
{N}=matrix of shell membrane forces;
$[N_1]$,$[N_2]$, $[N_3]$, $[N_4]$, $[N_5]$=shape function matrices;
P=point on the shell;
Q_{13},Q_{23}=shell shear forces;
Q_{11},Q_{22},Q_{33},Q_{12},Q_{44},Q_{55}=reduced stiffness material constants, (see equation (6.15));
{Q}=matrix of shear forces;
r=radial coordinate in the cross-section of a circular toroidal shell;
r,θ,z=cylindrical coordinates;
r,z=axisymmetric cylindrical coordinates;
R=radius of toroidal shell relative to the z axis;
R_1,R_2=radius of curvature for meridians, radius of curvature for parallel circles;
s=coordinate normal to the shell;
t=time;
t=space between one-dimensional finite element nodes, see Figure 3.3;
u_1,u_2,u_3=general shell displacements;
u,v,w=displacements corresponding with cylindrical r,θ,z coordinates;
u_θ,u_φ,u_3,β_θ,β_φ=displacements and rotations as defined in Section 3;
$\vec{u}$=displacement vector;
U_1,U_2,U_3,B_1,B_2=general displacements and rotations in axisymmetric system;
U,V,W,B_θ,B_φ=shell displacements and rotations as defined in Section 3;
U,V,W=elasticity displacements in axisymmetric system defined in Sections 4 and 5;
$\{u_1\}, \{u_2\}, \{u_3\}, \{\beta_1\}, \{\beta_2\}$=matrices of nodal point unknowns;

x,y,z=Cartesian coordinates;
α_1 ,α_2=shell meridian coordinate, shell parallel circle coordinate;
β_1,β_2=general shell rotations;
$\vec{\nabla}$=vector gradient operator;
ε_{11}^o, ε_{22}^o, ε_{12}^o=general shell membrane strains;
$\varepsilon_{\vartheta\vartheta}^o$, $\varepsilon_{\varphi\varphi}^o$, $\varepsilon_{\varphi\vartheta}^o$=shell membrane strains as defined in Section 3;
ε_{13}, ε_{23}=general shell shear strains;
$\varepsilon_{\vartheta 3}$, $\varepsilon_{\varphi 3}$=shell shear strains as defined in Section 3;
$\varepsilon_{rr}, \varepsilon_{\vartheta\vartheta}, \varepsilon_{\varphi\varphi}, \varepsilon_{\vartheta\varphi}, \varepsilon_{r\varphi}, \varepsilon_{r\vartheta}$=strains in the three-dimensional toroidal coordinate system;
$\varepsilon_{rr}, \varepsilon_{\vartheta\vartheta}, \varepsilon_{zz}, \varepsilon_{\vartheta z}, \varepsilon_{rz}, \varepsilon_{r\vartheta}$=strains in the three-dimensional cylindrical coordinate system;
$\varepsilon_1, \varepsilon_2, \varepsilon_3, \varepsilon_4, \varepsilon_5, \varepsilon_6$=strains written in matrix notation;
$\{\varepsilon\}$=strain matrix;
$\{\varepsilon^o\}$=matrix of shell membrane strains;
$\{\varepsilon_s\}$=matrix of shear strains;
$\bar{\bar{\varepsilon}}$=strain tensor;
κ_{11}, κ_{22}, κ_{12}=general shell bending curvatures;
$\kappa_{\vartheta\vartheta}$, $\kappa_{\varphi\varphi}$, $\kappa_{\vartheta\varphi}$=shell bending curvatures as defined in Section 3;
$\{\kappa\}$=matrix of shell bending curvatures;
δ,λ=nondimensional terms defined by equation (4.23);
ϑ, φ=toroidal coordinates;
$[\rho]$=mass density matrix;
$\sigma_{rr}, \sigma_{\vartheta\vartheta}, \sigma_{\varphi\varphi}, \sigma_{\vartheta\varphi}, \sigma_{r\varphi}, \sigma_{r\vartheta}$=stresses in the three-dimensional toroidal coordinate system;
$\sigma_{xx}, \sigma_{yy}, \sigma_{zz}, \sigma_{yz}, \sigma_{xz}, \sigma_{xy}$= stresses in a Cartesian coordinate system (see equation (6.1));
$\sigma_1, \sigma_2, \sigma_3, \sigma_4, \sigma_5, \sigma_6$=stresses written in matrix notation;
ν=Poisson's ratio;
ω=circular frequency;
Ω=nondimensional circular frequency;

1. INTRODUCTION

Vibration of shells is a topic that has received the attention of researchers for decades. Early work focused on analysis of the governing equations and exact solutions were dependent upon the ability of the analyst to obtain a solution in the coordinate system describing the shell. Finite difference methods gave researchers another computational tool. But the finite difference method was limited until the advent of computers that enabled an analysis that involved a solution for a significant number of equations. The finite element method gave researchers still another powerful computational tool for the analysis of general shells. The literature that pertains to shell theory is extensive. Toroidal shells have received some attention in the literature, but the coverage is limited when compared with cylindrical and spherical shells. The terminology in this Chapter when referring to thin shells implies that it will mean thin shells with shear deformation and rotary inertia unless clarified to be otherwise.

The current Chapter will focus on vibration of axisymmetric toroidal shells. The discussion will be based upon finite element analysis and in particular analysis that can be performed using a state-of-the-art personal computer (PC). Section 2 introduces the fundamental thin shell equations. Shell equations are not derived, but referenced to readily available textbooks that can dedicate many more pages to the topic than can be attempted here. There are three basic equations that relate membrane strain to in-plane forces, three equations that relate bending curvature to bending moments and two equations that relate transverse shear strain and shear forces. Material constants for isotropic engineering materials are set forth in matrix format for later use. A general finite element formulation is given for any thin shell analysis that includes shear deformation and rotary inertia.

Section 3 introduces the necessary theory to extend the equations of Section 2 to toroidal shells. The required Lamé parameters that relate the shell to its coordinate system are derived and the radii of curvature for toroidal shells are developed. A general finite element analysis for axisymmetric thin toroidal shells is derived in terms of the eight strains and curvatures of Section 2.

Results for frequency of free vibration for shells with free boundary conditions are computed. The analysis, as presented in this section, is a new contribution to the literature on vibration of toroidal shells.

Section 4 is dedicated to a three-dimensional elasticity analysis of toroidal shells. Three-dimensional elasticity equations are developed directly in the toroidal coordinate system. The advantage when using the equations of elasticity is that the thickness of the shell is not limited and thick shells, even solid shells, can be analyzed. Generalized equations of elasticity in dyadic format are specialized to describe a torus in a toroidal coordinate system. The finite element model is developed and toroidal shells of circular cross-section with moderately thick walls are analyzed.

The results of Section 3 are compared with the elasticity solution and the results of the thin shell analysis are verified. The analysis is extended to include moderately thick toroidal shells that have a sector of the cross-section removed.

Section 5 is dedicated to modeling a torus in an axisymmetric cylindrical coordinate system. The required finite element concepts are developed for an axisymmetric system. Toroidal shells with elliptical cross-section are analyzed and frequency of vibration and representative mode shapes are shown. A final application is an oblong cross-section with spherical caps. The intent is to demonstrate the powerful capability of the finite element method.

An introduction to toroidal shells with transversely isotropic material properties is presented in Section 6. The plane of isotropy is assumed to be at the surface of the shell such that the torus is transversely isotropic with respect to an axis normal to the shell. A second analysis assumes the torus is transversely isotropic with respect to the axis defining the circular direction around the torus. It is demonstrated that satisfactory results for frequency of vibration can be obtained using the thin shell formulation of Section 3 or the elasticity formulation of Section 4.

Again, the vibration analysis of transversely isotropic toroidal shells using the thin shell equations with shear deformation and rotary inertia represents a new contribution to the literature.

A final section is devoted to some conclusions concerning the intent of the Chapter and the results that have been presented.

2. General Finite Element Formulation for Thin Shells of Revolution Including Shear Deformation and Rotary Inertia

The equations that govern the behavior of thin shells have been recorded by numerous authors and for this Chapter the text by Soedel (1993) will serve as the primary source for equations that describe the behavior of thin shells of revolution. For such a shell, the lines of principal curvature are the meridians and the parallel circles. The coordinate for the meridian is, following Soedel (1993), α_1 and that for a parallel circle is α_2. The radius of curvature for the meridian is called R_1 and for the parallel circles is R_2. Fundamental to shell theory are the Lamé parameters which are equivalent to the scale factors that can be derived for the coordinate system that describes the shell. Section 3 of this chapter is devoted to describing these parameters for a toroidal shell of circular cross-section.

The shell strain-displacement equations are divided into membrane strains and corresponding in-plane forces; bending curvatures and corresponding bending moments; and for shear deformation the shear strains and corresponding shear forces. Note, if shear deformation is neglected the membrane and bending effects uncouple. In the case of analytical solutions the equations are simplified, but for finite element analysis the equations are easier to deal with if shear deformation is included.

The equations that describe in-plane forces, bending moments and transverse shears are not required for a free vibration analysis. In the case of a displacement analysis and stress analysis they would be required. But, that issue will be dealt with elsewhere in this text.

The general equations for in-plane strains, curvatures and shear strains are as follows where u_1, u_2 and u_3 are displacements, β_1 and β_2 are rotations, A_1 and A_2 are the Lamé parameters and R_1 and R_2 are the radii of curvature, all as defined above.

$$\epsilon_{11}^{o} = \frac{1}{A_1}\frac{\partial u_1}{\partial \alpha_1} + \frac{u_2}{A_1 A_2}\frac{\partial A_1}{\partial \alpha_2} + \frac{u_3}{R_1} \tag{2.1}$$

$$\epsilon_{22}^{o} = \frac{1}{A_2}\frac{\partial u_2}{\partial \alpha_2} + \frac{u_1}{A_1 A_2}\frac{\partial A_2}{\partial \alpha_1} + \frac{u_3}{R_2} \tag{2.2}$$

$$\epsilon_{12}^{o} = \frac{A_2}{A_1}\frac{\partial}{\partial \alpha_1}\left(\frac{u_2}{A_2}\right) + \frac{A_1}{A_2}\frac{\partial}{\partial \alpha_2}\left(\frac{u_1}{A_1}\right) \tag{2.3}$$

$$\kappa_{11} = \frac{1}{A_1}\frac{\partial \beta_1}{\partial \alpha_1} + \frac{\beta_2}{A_1 A_2}\frac{\partial A_1}{\partial \alpha_2} \tag{2.4}$$

$$\kappa_{22} = \frac{1}{A_2}\frac{\partial \beta_2}{\partial \alpha_2} + \frac{\beta_1}{A_1 A_2}\frac{\partial A_2}{\partial \alpha_1} \tag{2.5}$$

$$\kappa_{12} = \frac{A_2}{A_1}\frac{\partial}{\partial \alpha_1}\left(\frac{\beta_2}{A_2}\right) + \frac{A_1}{A_2}\frac{\partial}{\partial \alpha_2}\left(\frac{\beta_1}{A_1}\right) \tag{2.6}$$

$$\epsilon_{13} = \beta_1 - \frac{u_1}{R_1} + \frac{1}{A_1}\frac{\partial u_3}{\partial \alpha_1} \tag{2.7}$$

$$\epsilon_{23} = \beta_2 - \frac{u_2}{R_2} + \frac{1}{A_2}\frac{\partial u_3}{\partial \alpha_2} \tag{2.8}$$

Constitutive material constants connect the membrane forces and membrane strains as,

$$\{N\} = \begin{Bmatrix} N_{11} \\ N_{22} \\ N_{12} \end{Bmatrix} = [D_m]\{\epsilon^o\} = \begin{bmatrix} k_{11} & k_{12} & 0 \\ k_{12} & k_{11} & 0 \\ 0 & 0 & k_{33} \end{bmatrix} \begin{Bmatrix} \epsilon_{11}^o \\ \epsilon_{22}^o \\ \epsilon_{12}^o \end{Bmatrix} \tag{2.9}$$

Similarly, bending moments are defined in terms of curvatures and shear forces are defined in terms of shear strains,

$$\{M\} = \begin{Bmatrix} M_{11} \\ M_{22} \\ M_{12} \end{Bmatrix} = [D_b]\{\kappa\} = \begin{bmatrix} D_{11} & D_{12} & 0 \\ D_{12} & D_{22} & 0 \\ 0 & 0 & D_{33} \end{bmatrix} \begin{Bmatrix} \kappa_{11} \\ \kappa_{22} \\ \kappa_{12} \end{Bmatrix} \tag{2.10}$$

$$\{Q\} = \begin{Bmatrix} Q_{13} \\ Q_{23} \end{Bmatrix} = [D_s]\{\epsilon_s\} = \begin{bmatrix} G_{13} & 0 \\ 0 & G_{23} \end{bmatrix} \begin{Bmatrix} \epsilon_{13} \\ \epsilon_{23} \end{Bmatrix} \tag{2.11}$$

Material constants in equations (9) through (11) are as follows,

$$k_{11} = \frac{Eh}{(1-\nu^2)}, \qquad k_{12} = \nu k_{11}, k_{33} = k_{11}\frac{(1-\nu)}{2}$$

$$D_{11} = \frac{Eh^3}{12(1-\nu^2)}, D_{12} = \nu D_{11}, D_{33} = D_{11}\frac{(1-\nu)}{2} \tag{2.12}$$

$$G_{13} = G_{23} = k'Gh$$

It follows that the material constants are isotropic and k'(make sure that the text notation of k' matches the notation used in equation 2.12) is the shear correction factor and can be used as 2/3 or $\pi^2/12$.

The foregoing equations are sufficient to establish a finite element model for free vibration of thin shells of revolution. The fundamental finite element model can be developed using an energy method, such as Hamilton's principle as discussed by Meriovitch (1967), or the Galerkin method as presented by Reddy (1993). Either approach will lead to identical results. Assume that the five unknown displacements and rotations can be approximated in terms of shape functions as,

$$u_1 = [N_1]\{u_1\}, \qquad u_2 = [N_2]\{u_2\}, u_3 = [N_3]\{u_3\},$$

$$\beta_1 = [N_4]\{\beta_1\}, \qquad \beta_2 = [N_5]\{\beta_2\} \tag{2.13}$$

where $[N_i]$ is the shape function and the terms in { } are the nodal point unknowns. It follows that the same shape function can be used for all nodal unknowns. Equation (2.13) merely shows that different shape functions could be used for some variables. The free vibration analysis results in an eigenvalue analysis that is obtained by assuming harmonic motion defined as

$$u_1(\alpha_1, \alpha_2, s, t) = U_1(\alpha_1, \alpha_2, s)e^{i\omega t}$$

$$u_2(\alpha_1, \alpha_2, s, t) = U_2(\alpha_1, \alpha_2, s)e^{i\omega t}$$

$$u_3(\alpha_1, \alpha_2, s, t) = U_3(\alpha_1, \alpha_2, s)e^{i\omega t} \tag{2.14}$$

$$\beta_1(\alpha_1, \alpha_2, s, t) = \mathrm{B}_1(\alpha_1, \alpha_2, s)e^{i\omega t}$$

$$\beta_2(\alpha_1, \alpha_2, s, t) = \mathrm{B}_2(\alpha_1, \alpha_2, s)e^{i\omega t}$$

where s is the coordinate normal to the shell and ω is the circular frequency.

The free vibration problem is formulated in terms of a stiffness matrix [K] that results from the strain energy and a mass matrix [M] that results from the kinetic energy. It follows that

$$([K] - \omega^2[M])\{d\} = 0 \tag{2.15}$$

It follows from (2.14) that {d} is defined as

$$\{d\} = [U_1 \ \ U_2 \ \ U_3 \ \ B_1 \ \ B_2]^T \tag{2.16}$$

The stiffness matrix and mass matrix are defined in the traditional manner as,

$$[K] = \int_V [B]^T [\bar{D}] [B] dV \tag{2.17}$$

$$[M] = \int_V [N]^T [\rho] [N] dV \tag{2.18}$$

The stiffness matrix is traditionally defined in terms of a [B] matrix that is defined as

$$[B] = [L]\,[N] \tag{2.19}$$

[L] is an operator matrix and is dependent upon the partial differential equations that describe the physical problem. Application of the thin shells equations (2.1-2.8) dictates the [L] matrix as eight action type quantities dependent upon five displacement type variables, hence [L] is a 8x5 matrix. The shape function matrix has yet to be defined, but to satisfy equation (2.19) will have five rows. The material matrix $[\bar{D}]$ is obtained as an extension of the ideas reported by Tessler (1982) and results by combining equations (2.9-2.11) as

$$\begin{Bmatrix} N_{11} \\ N_{22} \\ N_{12} \\ M_{11} \\ M_{22} \\ M_{12} \\ Q_{13} \\ Q_{23} \end{Bmatrix} = \begin{bmatrix} k_{11} & k_{12} & 0 & 0 & 0 & 0 & 0 & 0 \\ k_{12} & k_{11} & 0 & 0 & 0 & 0 & 0 & 0 \\ 0 & 0 & k_{33} & 0 & 0 & 0 & 0 & 0 \\ 0 & 0 & 0 & D_{11} & D_{12} & 0 & 0 & 0 \\ 0 & 0 & 0 & D_{12} & D_{11} & 0 & 0 & 0 \\ 0 & 0 & 0 & 0 & 0 & D_{33} & 0 & 0 \\ 0 & 0 & 0 & 0 & 0 & 0 & G_{13} & 0 \\ 0 & 0 & 0 & 0 & 0 & 0 & 0 & G_{23} \end{bmatrix} \begin{Bmatrix} \epsilon_{11}^{o} \\ \epsilon_{22}^{o} \\ \epsilon_{12}^{o} \\ \kappa_{11} \\ \kappa_{22} \\ \kappa_{12} \\ \epsilon_{13} \\ \epsilon_{23} \end{Bmatrix} \tag{2.20}$$

The material matrix is 8x8 and remains the same for application to the analysis of any axisymmetric thin shell. The matrix [ρ] is the complete counterpart of the mass density definition given by Tessler (1982) and is similar to the definitions that are used in first order shear deformation plate theory. The displacement mass term is ρh and rotary inertia term is $\rho h^3/12$ as follows,

$$[\rho] = \begin{bmatrix} \rho h & 0 & 0 & 0 & 0 \\ 0 & \rho h & 0 & 0 & 0 \\ 0 & 0 & \rho h & 0 & 0 \\ 0 & 0 & 0 & \rho h^3/12 & 0 \\ 0 & 0 & 0 & 0 & \rho h^3/12 \end{bmatrix} \tag{2.21}$$

It follows that equation (2.21) is independent of the shell geometry.

The operator matrix [L] is defined as a general statement based upon equations (2.1-2.8) as,

$$[L] = \begin{bmatrix} \frac{1}{A_1}\frac{\partial}{\partial \alpha_1} & \frac{1}{A_1 A_2}\frac{\partial A_1}{\partial \alpha_2} & \frac{1}{R_1} & 0 & 0 \\ \frac{1}{A_1 A_2}\frac{\partial A_2}{\partial \alpha_1} & \frac{1}{A_2}\frac{\partial}{\partial \alpha_2} & \frac{1}{R_2} & 0 & 0 \\ \frac{A_1}{A_2}\frac{\partial}{\partial \alpha_2}\left(\frac{}{A_1}\right) & \frac{A_2}{A_1}\frac{\partial}{\partial \alpha_1}\left(\frac{}{A_2}\right) & 0 & 0 & 0 \\ 0 & 0 & 0 & \frac{1}{A_1}\frac{\partial}{\partial \alpha_1} & \frac{1}{A_1 A_2}\frac{\partial A_1}{\partial \alpha_2} \\ 0 & 0 & 0 & \frac{1}{A_1 A_2}\frac{\partial A_2}{\partial \alpha_1} & \frac{1}{A_2}\frac{\partial}{\partial \alpha_2} \\ 0 & 0 & 0 & \frac{A_1}{A_2}\frac{\partial}{\partial \alpha_2}\left(\frac{}{A_1}\right) & \frac{A_2}{A_1}\frac{\partial}{\partial \alpha_1}\left(\frac{}{A_2}\right) \\ -\frac{1}{R_1} & 0 & \frac{1}{A_1}\frac{\partial}{\partial \alpha_1} & 1 & 0 \\ 0 & -\frac{1}{R_2} & \frac{1}{A_2}\frac{\partial}{\partial \alpha_2} & 0 & 1 \end{bmatrix} \tag{2.22}$$

Equation (2.22) can be used to study a two-dimensional thin shell because the assumptions that render the shell axisymmetric have yet to be incorporated in the analysis. Thus far, the analysis is completely general. The next step would be to choose the shell geometry and decide if the analysis will be axisymmetric. Once the coordinate system has been determined the coordinates α_1 and α_2 will be known. The Lamé parameters A_1 and A_2 can be computed and the radii R_1 and R_2 can be identified. The last step would be to decide upon a shape function [N]. If the geometry is two-dimensional that would dictate a two-dimensional shape function, but if axisymmetric it would only require a one-dimensional shape function.

3. Finite Element Formulation for Free Vibration of Thin Toroidal Shells with Circular Cross-Section Including Shear Deformation and Rotary Inertia

The concept developed in the preceding section will be specialized to toroidal coordinates. The toroidal coordinate system (r,θ,φ) is shown in Figure 3.1 with shell coordinates

$$\alpha_1 = \vartheta \text{ and } \alpha_2 = \varphi \tag{3.1}$$

The coordinate system has been used by several authors, Zhang and Redekop (1992), Ming, Pan and Norton (2002) and Buchanan and Liu (2005) among others. The coordinate angle θ is measured from the horizontal axis on the shell cross-section. Angular measure φ about the z axis is measured from the y axis.

The Lamé parameters are usually obtained in a standard form that is developed along with thin shell theory. In this Chapter thin shell theory has been referenced to Soedel (1993) as opposed to being derived. Rather than reference the Lamé parameters to thin shell theory, they will be developed using the scale factors that define the toroidal coordinate system. Scale factors will be required for the theory of Section 4. Lamé parameters are merely the scale factors with the r coordinate replaced with the radius of the thin toroidal shell. The derivation of scale factors follows the concepts presented by Spiegel (1959).

The coordinates of a point P of Figure 3.1 are,

$$x = (R + r\cos\vartheta)\sin\varphi \tag{3.2}$$

$$y = (R + r\cos\vartheta)\cos\varphi \tag{3.3}$$

$$z = r\sin\vartheta \tag{3.4}$$

where R is the radius of the torus and r is a coordinate that locates a point inside the torus. The radius a defines the radius of the thin shell measured to the mid-plane of the thin shell. The position vector that defines the point P is given by,

$$\vec{P} = (R + r\cos\vartheta)\sin\varphi\vec{i} + (R + r\cos\vartheta)\cos\varphi\vec{j} + r\sin\vartheta\vec{k} \tag{3.5}$$

Also, $d\vec{P} = \frac{\partial \vec{P}}{\partial r}dr + \frac{\partial \vec{P}}{\partial \vartheta}d\vartheta + \frac{\partial \vec{P}}{\partial \varphi}d\varphi$

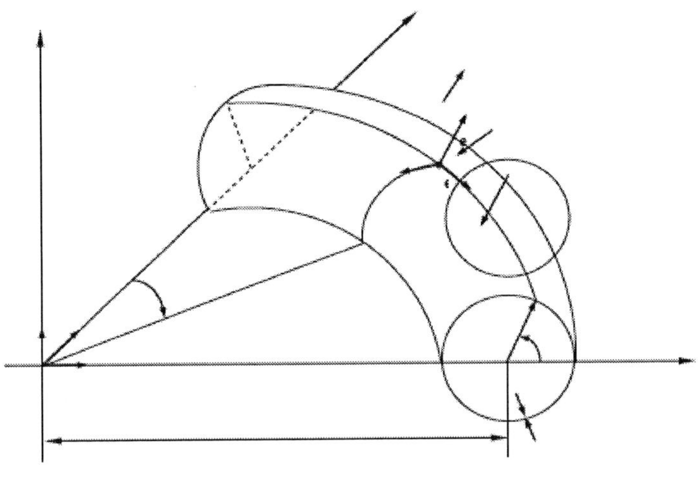

Figure 3.1. Toroidal coordinate system and *x-z* or *y-z* plane cross-section view.

Scale factors are computed as

$$h_r = \left|\frac{\partial P}{\partial r}\right| = \left|\frac{\partial \vec{P}}{\partial r}\cdot\frac{\partial \vec{P}}{\partial r}\right|^{\frac{1}{2}} = 1 \tag{3.6}$$

$$h_\vartheta = \left|\frac{\partial P}{\partial \vartheta}\right| = \left|\frac{\partial \vec{P}}{\partial \vartheta}\cdot\frac{\partial \vec{P}}{\partial \vartheta}\right|^{\frac{1}{2}} = r \tag{3.7}$$

$$h_\varphi = \left|\frac{\partial P}{\partial \varphi}\right| = \left|\frac{\partial \vec{P}}{\partial \varphi}\cdot\frac{\partial \vec{P}}{\partial \varphi}\right|^{\frac{1}{2}} = R + rcos\vartheta \tag{3.8}$$

Equations (3.6-3.8) will be used in the next section, but for Lamé parameters r becomes the thin shell radius a. Then,

$$A_1 = a, \qquad A_2 = R + acos\vartheta \tag{3.9}$$

The shell radii R_1 and R_2 are shown in Figure 3.2.

$$R_1 = a, \qquad R_2 = (R + acos\vartheta)/cos\vartheta \tag{3.10}$$

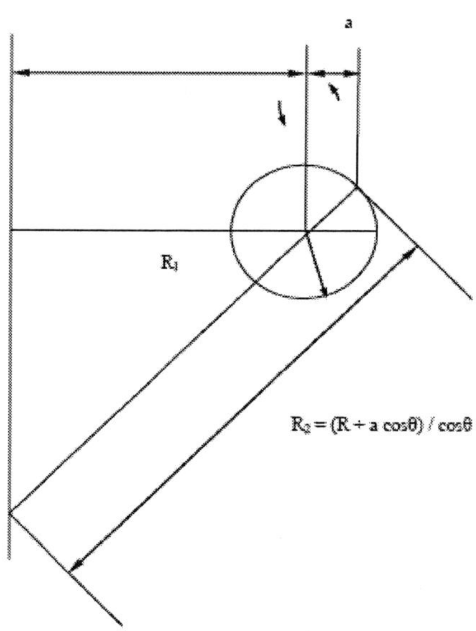

Figure 3.2. View of the toroidal shell showing radii of curvature, R_1 and R_2.

Displacements become u_θ, u_φ and u_3 (u_3 is the notation used by Soedel (1993)) with rotations given as β_θ and β_φ. The strain displacement equations and curvature rotation equations are obtained by substituting equations (3.1), (3.9) and (3.10) into equations (2.1) through (2.8) to give,

$$\epsilon_{\vartheta\vartheta}^{o} = \frac{1}{a}\frac{\partial u_\vartheta}{\partial \vartheta} + \frac{u_3}{a} \tag{3.11}$$

$$\epsilon^{o}_{\varphi\varphi} = -\frac{u_{\vartheta} sin\vartheta}{R+a\ cos\vartheta} + \frac{1}{R+a\ cos\vartheta}\frac{\partial u_{\varphi}}{\partial\varphi} + \frac{u_3 cos\theta}{R+a\ cos\vartheta} \tag{3.12}$$

$$\epsilon^{o}_{\phi\theta} = \frac{1}{R+a\ cos\vartheta}\frac{\partial u_{\vartheta}}{\partial\varphi} + \frac{1}{a}\frac{\partial u_{\varphi}}{\partial\vartheta} + \frac{u_{\varphi} sin\phi}{R+a\ cos\vartheta} \tag{3.13}$$

$$\kappa_{\vartheta\vartheta} = \frac{1}{a}\frac{\partial\beta_{\vartheta}}{\partial\vartheta} \tag{3.14}$$

$$\kappa_{\varphi\varphi} = -\frac{\beta_{\vartheta} sin\vartheta}{R+a\ cos\vartheta} + \frac{1}{R+a\ cos\vartheta}\frac{\partial\beta_{\varphi}}{\partial\varphi} \tag{3.15}$$

$$\kappa_{\vartheta\varphi} = \frac{1}{R+a\ cos\vartheta}\frac{\partial\beta_{\vartheta}}{\partial\varphi} + \frac{1}{a}\frac{\partial\beta_{\varphi}}{\partial\vartheta} + \frac{\beta_{\varphi} sin\vartheta}{R+a\ cos\vartheta} \tag{3.16}$$

$$\epsilon_{\vartheta 3} = -\frac{u_{\vartheta}}{a} + \frac{1}{a}\frac{\partial u_3}{\partial\vartheta} + \beta_{\vartheta} \tag{3.17}$$

$$\epsilon_{\varphi 3} = -\frac{u_{\varphi} cos\vartheta}{R+a\ cos\vartheta} + \frac{1}{R+a\ cos\vartheta}\frac{\partial u_3}{\partial\varphi} + \beta_{\varphi} \tag{3.18}$$

A finite element analysis for free vibration of a thin toroidal shell including shear deformation and rotary inertia is based upon equations (3.11) through (3.18). The free vibration problem was defined in the preceding section by equation (2.15) and is repeated below.

$$([K] - \omega^2[M])\{d\} = 0 \tag{2.15}$$

The stiffness matrix [K] was previously defined as,

$$[K] = \int_V [B]^T\,[\bar{D}]\,[B]dV \tag{2.17}$$

And the [B] matrix was previously defined as,

$$[B] = [L]\,[N] \tag{2.19}$$

The operator matrix [L] can define either a two-dimensional segment of the shell in (θ, φ) coordinates or an axisymmetric shell in terms of the ϑ coordinate. The two-dimensional operator [L] is based upon equation (2.22) as follows.

$$[L] = \begin{bmatrix} \frac{1}{a}\frac{\partial}{\partial\vartheta} & 0 & \frac{1}{a} & 0 & 0 \\ -\frac{sin\vartheta}{R+acos\vartheta} & \frac{1}{R+acos\vartheta}\frac{\partial}{\partial\varphi} & \frac{cos\vartheta}{R+acos\vartheta} & 0 & 0 \\ \frac{1}{R+acos\vartheta}\frac{\partial}{\partial\varphi} & \frac{1}{a}\frac{\partial}{\partial\vartheta}+\frac{sin\vartheta}{R+acos\vartheta} & 0 & 0 & 0 \\ 0 & 0 & 0 & \frac{1}{a}\frac{\partial}{\partial\vartheta} & 0 \\ 0 & 0 & 0 & -\frac{sin\vartheta}{R+acos\vartheta} & \frac{1}{R+acos\vartheta}\frac{\partial}{\partial\varphi} \\ 0 & 0 & 0 & \frac{1}{R+acos\vartheta}\frac{\partial}{\partial\varphi} & \frac{1}{a}\frac{\partial}{\partial\vartheta}+\frac{sin\vartheta}{R+acos\vartheta} \\ -\frac{1}{a} & 0 & \frac{1}{a}\frac{\partial}{\partial\vartheta} & 1 & 0 \\ 0 & -\frac{cos\vartheta}{R+acos\vartheta} & \frac{1}{R+acos\vartheta}\frac{\partial}{\partial\varphi} & 0 & 1 \end{bmatrix} \tag{3.19}$$

The [B] matrix could be completed by assuming a suitable two-dimensional shape function and proceeding to equation (2.17). In keeping with the subject of this chapter the axisymmetric analysis will be discussed in some detail.

Similar to equation (2.14), assume a solution that renders the analysis axisymmetric with harmonic motion.

$$u_\theta(\vartheta,\varphi,r,t) = U(\vartheta,r)\cos n\varphi\, e^{i\omega t}$$

$$u_\phi(\vartheta,\varphi,r,t) = V(\vartheta,r)\sin n\varphi\, e^{i\omega t}$$

$$u_3(\vartheta,\varphi,r,t) = W(\vartheta,r)\cos n\varphi\, e^{i\omega t} \qquad (3.20)$$

$$\beta_\vartheta(\theta,\varphi,r,t) = B_\vartheta(\vartheta,r)\cos n\varphi\, e^{i\omega t}$$

$$\beta_\varphi(\vartheta,\varphi,r,t) = B_\varphi(\vartheta,r)\sin n\varphi\, e^{i\omega t}$$

where n is a circumferential wave number and ω is the circular frequency. Note, some authors prefer to use cosωt or sinωt rather than $e^{i\omega t}$ to establish free vibration. The end result is the same. Also, the coordinate normal to the shell was used as s (following Soedel (1993)) in equation (2.14) and has been replaced with r in equation (3.20).

Assume a quadratic one-dimensional shape function. The shape function can be either isoparametric or written directly in terms of ϑ. In either case Gauss-Legendre quadrature is used for evaluating the volume integrals. A quadratic shape function has the following matrix representation

$$[N] = \begin{bmatrix} N_1 & 0 & 0 & 0 & 0 & N_2 & 0 & 0 & 0 & 0 & N_3 & 0 & 0 & 0 & 0 \\ 0 & N_1 & 0 & 0 & 0 & 0 & N_2 & 0 & 0 & 0 & 0 & N_3 & 0 & 0 & 0 \\ 0 & 0 & N_1 & 0 & 0 & 0 & 0 & N_2 & 0 & 0 & 0 & 0 & N_3 & 0 & 0 \\ 0 & 0 & 0 & N_1 & 0 & 0 & 0 & 0 & N_2 & 0 & 0 & 0 & 0 & N_3 & 0 \\ 0 & 0 & 0 & 0 & N_1 & 0 & 0 & 0 & 0 & N_2 & 0 & 0 & 0 & 0 & N_3 \end{bmatrix} \qquad (3.21)$$

Substitute equations (3.20) and (3.21) into equation (3.19) to obtain the [B] matrix as [B]=[L][N]. The [B] matrix for an axisymmetric formulation follows as,

$$[B] = \begin{bmatrix} \frac{1}{a}\frac{\partial N_1}{\partial\vartheta} & 0 & \frac{N_1}{a} & 0 & 0 & \cdots \\ -\frac{N_1 sin\vartheta}{R+acos\vartheta} & \frac{nN_1}{R+acos\vartheta} & \frac{N_1 cos\vartheta}{R+acos\vartheta} & 0 & 0 & \cdots \\ -\frac{nN_1}{R+acos\vartheta} & \frac{1}{a}\frac{\partial N_1}{\partial\vartheta}+\frac{N_1 sin\vartheta}{R+acos\vartheta} & 0 & 0 & 0 & \cdots \\ 0 & 0 & 0 & \frac{1}{a}\frac{\partial N_1}{\partial\vartheta} & 0 & \cdots \\ 0 & 0 & 0 & -\frac{N_1 sin\vartheta}{R+acos\vartheta} & \frac{nN_1}{R+acos\vartheta} & \cdots \\ 0 & 0 & 0 & -\frac{nN_1}{R+acos\vartheta} & \frac{1}{a}\frac{\partial N_1}{\partial\vartheta}+\frac{N_1 sin\vartheta}{R+acos\vartheta} & \cdots \\ -\frac{N_1}{a} & 0 & \frac{1}{a}\frac{\partial N_1}{\partial\vartheta} & N_1 & 0 & \cdots \\ 0 & -\frac{N_1 cos\vartheta}{R+acos\vartheta} & -\frac{nN_1}{R+acos\vartheta} & 0 & N_1 & \cdots \end{bmatrix} \qquad (3.22)$$

There are ten additional columns corresponding to N_2 and N_3. The material matrix is identical to equation (2.20) and the mass matrix is the same as equation (2.21). The {d} matrix of equation (2.17) corresponds to equations (3.20) as

$$\{d\} = \left[U\ V\ W\ B_{\vartheta}\ B_{\varphi}\right] \tag{3.23}$$

Finally, the volume integral of equation (2.17) is completed by defining dV. It follows that dV can be defined in terms of the scale factors as,

$$dV = h_{\varphi}\vec{e}_{\varphi}h_{\vartheta}\vec{e}_{\vartheta}h_r\vec{e}_r d\varphi d\vartheta dr \tag{3.24}$$

For a two-dimensional formulation dV becomes

$$dV = a(R + acos\vartheta)hd\vartheta d\varphi \tag{3.25}$$

And for the axisymmetric analysis corresponding to equation (3.22)

$$dV = 2\pi a(R + acos\vartheta)hd\vartheta \tag{3.25}$$

where h is the thickness of the shell and a is the distance to the mid-plane of the shell.

The axisymmetric formulation described in this section has successfully been used by Al-Khatib and Buchanan (2010) for free vibration of a paraboloidal shell. The analysis of a toroidal shell based upon a formulation that involves equation (3.22) will incorporate the same finite element that was used in Al-Khatib and Buchanan (2010). A one-dimensional quadratic three node Lagrangian element in terms of ϑ shown in Figure 3.3 can be written as,

$$N_1 = \frac{(\vartheta - t)(\vartheta - 2t)}{2t^2}, \quad N_2 = \frac{\vartheta(2t - \vartheta)}{t^2}, \quad N_3 = \frac{\vartheta(\vartheta - t)}{2t^2} \tag{3.26}$$

where t is the spacing between finite element nodes. Equation (3.26) is substituted into equation (3.21). Numerical integration is based upon the Gauss-Legendre quadrature and requires that the integration limits be changed from 0 to 2t to -1 to +1. Alternatively, equation (3.26) could be written as an isoparametric element, but a one-dimensional formulation would not require that degree of sophistication.

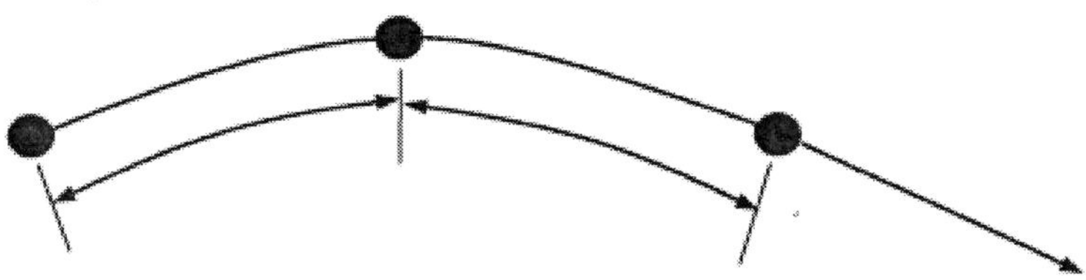

Figure 3.3. Quadratic one-dimensional finite element.

Results for frequency are given in terms of nondimensional parameters with outside radius of the shell a_o equal to unity along with density ρ and shear modulus G (see equation

(2.12)) equal to unity. It follows that $a_o = a + h/2 = 1$, where, as shown in Figure 2.1, a is the distance to the mid-plane of the shell and h is the thickness of the shell computed as a fraction of the outside radius of the shell. Material constants are computed using equations (2.12) with G=1.0, Poisson's ratio ν=0.3 giving E=2.6. Frequencies of free vibration for free boundary conditions were computed for five values of shell thickness h. The relation between a_o, a and h is shown in Table 3.1. Results for frequency were computed for three representative values of toroidal radius R as $2a_o$, $5a_o$ and $10a_o$. Nondimensional frequency is given in terms of the assumed nondimensional parameters as

$$\Omega = \omega a_o \sqrt{\rho/G} \tag{3.27}$$

Results are given in Tables 3.2 through 3.4. Note, these results for thin toroidal shells including shear deformation and rotary inertia computed using the concepts in this section are new and have not been published elsewhere.

Table 3.1. Parameters to compute $a_o = a(mid - plane) + h/2$

h	a	h/2	a_o
$0.2a_o$	0.900	0.100	1.0
$0.1a_o$	0.950	0.050	1.0
$0.05a_o$	0.975	0.025	1.0
$0.02a_o$	0.990	0.010	1.0
$0.01a_o$	0.995	0.005	1.0

Table 3.2. Frequencies $\Omega = \omega a_o \sqrt{\rho/G}$ for toroidal shells with free boundary conditions, $R = 2a_o$, ν=0.3 and $a_o = a(mid - plane) + h/2$

	h				
n	$0.2a_o$	$0.1a_o$	$0.05a_o$	$0.02a_o$	$0.01a_o$
0	0.2631 (1)	0.1604 (1)	0.1063 (1)	0.0643 (1)	0.0446 (1)
	0.5528 (8)	0.4642 (8)	0.4201 (10)	0.3815	0.3411
	0.8094	0.6228	0.4954	0.4093	0.3501
1	0.5034 (5)	0.4047 (4)	0.3672 (6)	0.3356 (8)	0.3098
	0.5307 (7)	0.4291 (7)	0.3899 (8)	0.3543	0.3221
	0.7630	0.6497	0.4924	0.4017	0.3438
2	0.3290 (2)	0.2573 (2)	0.1782 (2)	0.1047 (2)	0.0709 (2)
	0.3990 (3)	0.3038 (3)	0.2000 (3)	0.1143 (3)	0.0764 (3)
	0.4902 (4)	0.4106 (5)	0.3894 (7)	0.3452 (9)	0.3054
	0.5184 (6)	0.4241 (6)	0.3948 (9)	0.3518 (10)	0.3079
3	0.6811 (9)	0.5401 (9)	0.3619 (4)	0.1930 (4)	0.1219 (4)
	0.7010 (10)	0.5469 (10)	0.3639 (5)	0.1936 (5)	0.1223 (5)
	0.7533	0.6417	0.5129	0.3685	0.2972 (10)
4	0.9173	0.7663	0.5564	0.2936 (6)	0.1760 (6)
	0.9292	0.7672	0.5566	0.2936 (7)	0.1761 (7)
5	1.1327	0.9105	0.7328	0.4126	0.2420 (8)
	1.1399	0.9112	0.7328	0.4126	0.2420 (9)
6	1.3295	1.0480	0.8395	0.5398	0.3203
	1.3345	1.0486	0.8395	0.5398	0.3203

The first ten frequencies are indentified with numbers in parentheses. Sufficient frequency results are recorded to illustrate the behavior of the vibration as the radius of the torus is increased. The lowest frequency corresponds to n=0 for a lesser torus radius as shown in Table 3.2. As the radius increases the first frequency shifts to correspond to n=2, a fact that could be important in the design of toroidal shells.

Table 3.3. Frequencies $\Omega = \omega a_o\sqrt{\rho/G}$ for toroidal shells with free boundary conditions, $R = 5a_o$, v=0.3 and $a_o = a(mid - plane) + h/2$

n	h 0.2	0.1	0.05	0.02	0.01
0	0.1790 (3) 0.3155 (7) 0.4038	0.1102 (3) 0.2560 0.3361	0.0677 (3) 0.2217 0.2856	0.0403 (3) 0.1980 0.2180	0.0277 (3) 0.1788 0.1862
1	0.2619 (6) 0.3808 0.4110	0.1868 (6) 0.2603 0.3502	0.1514 (6) 0.2219 0.2925	0.1341 (10) 0.1988 0.2184	0.1265 0.1789 0.1860
2	0.0842 (1) 0.0873 (2) 0.3593 (9) 0.3739	0.0689 (1) 0.0698 (2) 0.2316 (7) 0.2458 (10)	0.0489 (1) 0.0492 (2) 0.1948 (9) 0.2105 (10)	0.0303 (1) 0.0313 (2) 0.1772 0.1916	0.0205 (1) 0.0215 (2) 0.1653 0.1749
3	0.2119 (4) 0.2217 (5) 0.3843	0.1613 (4) 0.1655 (5) 0.2803	0.1075 (4) 0.1092 (5) 0.2631	0.0625 (4) 0.0629 (5) 0.2128	0.0417 (4) 0.0427 (5) 0.1796
4	0.3539 (8) 0.3669 (10)	0.2391 (8) 0.2410 (9)	0.1612 (7) 0.1620 (8)	0.0885 (6) 0.0889 (7)	0.0573 (6) 0.0579 (7)
5	0.4654 0.4681	0.3182 0.3189	0.2188 0.2190	0.1172 (8) 0.1172 (9)	0.0735 (8) 0.0739 (9)
6	0.5541 0.5541	0.4072 0.4074	0.2772 0.2773	0.1485 0.1486	0.0912 (10) 0.0914

Table 3.4. Frequencies $\Omega = \omega a_o\sqrt{\rho/G}$ for toroidal shells with free boundary conditions, $R = 10a_o$, v=0.3 and $a_o = a(mid - plane) + h/2$

n	h 0.2	0.1	0.05	0.02	0.01
0	0.1072 (5) 0.1616 (9) 0.3330	0.0849 (5) 0.1592 0.1867	0.0526 (5) 0.1307 0.1646	0.0296 (5) 0.1122 0.1338	0.0202 (5) 0.1035 0.1114
1	0.1461 (8) 0.2262 0.3327	0.1227 (8) 0.1827 0.2029	0.0890 (8) 0.1316 0.1798	0.0710 0.1125 0.1344	0.0658 0.1038 0.1115
2	0.0256 (1) 0.0261 (2) 0.2267 0.3326	0.0214 (1) 0.0237 (2) 0.1674 0.1812	0.0174 (1) 0.0181 (2) 0.1162 0.1291	0.0108 (1) 0.0122 (2) 0.0974 0.1109	0.00780(1) 0.00945(2) 0.0908 0.1027
3	0.0684 (3) 0.0698 (4) 0.3126	0.0622 (3) 0.0628 (4) 0.1787	0.0462 (3) 0.0463 (4) 0.1311	0.0273 (3) 0.0281 (4) 0.1160	0.0185 (3) 0.0189 (4) 0.1060
4	0.1238 (6) 0.1263 (7)	0.1100 (6) 0.1110 (7)	0.0748 (6) 0.0750 (7)	0.0426 (6) 0.0429 (7)	0.0280 (6) 0.0288 (7)
5	0.1884 (10) 0.1914	0.1537 (9) 0.1546 (10)	0.0993 (9) 0.0996(10)	0.0567 (8) 0.0568 (9)	0.0362 (8) 0.0367 (9)
6	0.2580 0.2619	0.1861 0.1865	0.1248 0.1250	0.0709 (10) 0.0709	0.0447 (10) 0.0450

Also, the higher frequencies correspond to higher circumferential modes as the radius of the torus increases. A relative thick shell is represented by h as $0.2a_o$ and $0.1a_o$ and will be compared with a three-dimensional analysis in the next section.

4. Three-Dimensional Elasticity Equations and Free Vibration Analysis in a Toroidal Coordinate System

The three-dimensional elasticity equations governing the behavior of toroidal shells can be written in the toroidal coordinate system that was introduced in the previous section. The strain-displacement equations appear in the literature with application to both static and dynamic, Jiang and Redekop (2002), and dynamic analysis presented by Zhou et al. (2002). The concepts presented here are a more in-depth discussion of work presented by Buchanan and Liu (2005).

It is instructive to demonstrate the derivation of the strain-displacement equations directly from theory of elasticity. The development that follows is outlined in detail by Chow and Pagano (1967) and is based upon the equations of elasticity in dyadic notation. The dyadic form of strain-displacement is,

$$\bar{\bar{\varepsilon}} = \frac{1}{2}\left[\vec{\nabla}\vec{u} + \left(\vec{\nabla}\vec{u}\right)^{T}\right] \tag{4.1}$$

where $\bar{\bar{\varepsilon}}$ is the strain tensor and $\vec{u}$ follows from Figure 3.1 as,

$$\vec{u} = u_r\vec{e}_r + u_\vartheta\vec{e}_\theta + u_\varphi\vec{e}_\varphi \tag{4.2}$$

Recall the scale factors from Section 3,

$$h_r = 1, \qquad h_\vartheta = r, h_\varphi = R + r\cos\vartheta \tag{4.3}$$

The scale factors and the definition of the gradient operator as recorded by Hughes and Gaylord (1964) give $\vec{\nabla}$ in toroidal coordinates

$$\vec{\nabla} = \frac{\partial}{\partial r}\vec{e}_r + \frac{1}{r}\frac{\partial}{\partial\vartheta}\vec{e}_\vartheta + \frac{1}{R + r\cos\vartheta}\frac{\partial}{\partial\varphi}\vec{e}_\varphi \tag{4.4}$$

The derivatives of the unit vectors follow from the discussion in Chow and Pagano (1967) as,

$$\begin{pmatrix} \frac{\partial\vec{e}_r}{\partial r} & \frac{\partial\vec{e}_r}{\partial\vartheta} & \frac{\partial\vec{e}_r}{\partial\varphi} \\ \frac{\partial\vec{e}_\vartheta}{\partial r} & \frac{\partial\vec{e}_\vartheta}{\partial\vartheta} & \frac{\partial\vec{e}_\vartheta}{\partial\varphi} \\ \frac{\partial\vec{e}_\varphi}{\partial r} & \frac{\partial\vec{e}_\varphi}{\partial\vartheta} & \frac{\partial\vec{e}_\varphi}{\partial\varphi} \end{pmatrix} = \begin{pmatrix} 0 & \vec{e}_\vartheta & \vec{e}_\varphi\cos\vartheta \\ 0 & -\vec{e}_r & -\vec{e}_\varphi\sin\vartheta \\ 0 & 0 & \vec{e}_\vartheta\sin\vartheta - \vec{e}_r\cos\vartheta \end{pmatrix} \tag{4.5}$$

Tensor notation that is used in elasticity should be changed to a corresponding matrix notation where the subscripts have the following correspondence. Every four subscripted tensor quantity, such as the material tensor, can be converted to a two subscripted quantity using the following relationship.

tensor	11	22	33	23	13	12
	rr	θθ	φφ	θφ	rφ	rθ
matrix	1	2	3	4	5	6

It follows that the strain tensor is converted to matrix notation as follows.

$$\begin{pmatrix} \varepsilon_{rr} & \varepsilon_{r\vartheta} & \varepsilon_{r\varphi} \\ \varepsilon_{r\vartheta} & \varepsilon_{\vartheta\vartheta} & \varepsilon_{\vartheta\varphi} \\ \varepsilon_{r\varphi} & \varepsilon_{\theta\varphi} & \varepsilon_{\varphi\varphi} \end{pmatrix} = \begin{pmatrix} \varepsilon_1 & \varepsilon_6 & \varepsilon_5 \\ \varepsilon_6 & \varepsilon_2 & \varepsilon_4 \\ \varepsilon_5 & \varepsilon_4 & \varepsilon_3 \end{pmatrix} \tag{4.6}$$

Note, there is no accepted order for the correspondence of the matrix subscripts 4, 5, 6. The reader/researcher must be aware of the intention of the author.

The strain-displacement equations in toroidal coordinates based upon equation (4.1) are,

$$\varepsilon_{rr} = \varepsilon_1 = \frac{\partial u_r}{\partial r} \tag{4.7}$$

$$\varepsilon_{\vartheta\vartheta} = \varepsilon_2 = \frac{1}{r}\frac{\partial u_\vartheta}{\partial \vartheta} + \frac{u_r}{r} \tag{4.8}$$

$$\varepsilon_{\varphi\varphi} = \varepsilon_3 = \frac{1}{R + r cos\vartheta}\left(\frac{\partial u_\varphi}{\partial \varphi} + u_r cos\vartheta - u_\vartheta sin\vartheta\right) \tag{4.9}$$

$$\varepsilon_{\vartheta\varphi} = \varepsilon_4 = \frac{1}{r}\frac{\partial u_\varphi}{\partial \vartheta} + \frac{1}{R + r cos\vartheta}\left(\frac{\partial u_\vartheta}{\partial \varphi} + u_\varphi sin\vartheta\right) \tag{4.10}$$

$$\varepsilon_{r\varphi} = \varepsilon_5 = \frac{\partial u_\varphi}{\partial r} + \frac{1}{R + r cos\vartheta}\left(\frac{\partial u_r}{\partial \varphi} - u_\varphi cos\vartheta\right) \tag{4.11}$$

$$\varepsilon_{r\vartheta} = \varepsilon_6 = \frac{\partial u_\vartheta}{\partial r} - \frac{u_\vartheta}{r} + \frac{1}{r}\frac{\partial u_r}{\partial \vartheta} \tag{4.12}$$

Stress-strain relations for isotropic materials using matrix notation of this section are,

$$\begin{Bmatrix} \sigma_{rr} \\ \sigma_{\vartheta\vartheta} \\ \sigma_{\varphi\varphi} \\ \sigma_{\theta\varphi} \\ \sigma_{r\varphi} \\ \sigma_{r\vartheta} \end{Bmatrix} = \begin{Bmatrix} \sigma_1 \\ \sigma_2 \\ \sigma_3 \\ \sigma_4 \\ \sigma_5 \\ \sigma_6 \end{Bmatrix} = [C]\{\varepsilon\} = \begin{bmatrix} C_{11} & C_{12} & C_{12} & 0 & 0 & 0 \\ C_{12} & C_{11} & C_{12} & 0 & 0 & 0 \\ C_{12} & C_{12} & C_{11} & 0 & 0 & 0 \\ 0 & 0 & 0 & C_{44} & 0 & 0 \\ 0 & 0 & 0 & 0 & C_{44} & 0 \\ 0 & 0 & 0 & 0 & 0 & C_{44} \end{bmatrix} \begin{Bmatrix} \varepsilon_1 \\ \varepsilon_2 \\ \varepsilon_3 \\ \varepsilon_4 \\ \varepsilon_5 \\ \varepsilon_6 \end{Bmatrix} \tag{4.13}$$

Note, for isotropic materials $C_{44} = (C_{11} - C_{12})/2$.

The toroidal coordinate system lends itself to an axisymmetric formulation. The governing equations allow suitable assumptions that can reduce the finite element formulation from three dimensions to two dimensions and maintain the three-dimensional character of the shell. Solutions that satisfy the governing equations in terms of circumferential displacement φ and circular frequency ω are

$$u_r(r,\vartheta,\varphi,t) = U(r,\vartheta)cosn\varphi cos\omega t \tag{4.14}$$

$$u_\varphi(r,\vartheta,\varphi,t) = V(r,\vartheta)cosn\varphi cos\omega t \tag{4.15}$$

$$u_\vartheta(r,\vartheta,\varphi,t) = W(r,\vartheta)sinn\varphi cos\omega t \tag{4.16}$$

where n is the circumferential wave number and ω is the circular frequency. Zhou et al. (2002) used a similar assumption to simplify their analysis of solid toroidal shells.

As in the previous sections, free vibration analysis is defined in terms of the stiffness matrix and mass matrix.

$$([K] - \omega^2[M])\{d\} = 0 \tag{4.17}$$

The stiffness matrix [K] is defined as,

$$[K] = \int_V [B]^T\,[C]\,[B]dV \tag{4.18}$$

The material matrix [C] is based upon equation (4.13) and the [B] matrix is,

$$[B] = [L]\,[N] \tag{4.19}$$

The operator matrix [L] is obtained in terms of the strain-displacement equations (4.7) through (4.12) and [N] is a suitable two-dimensional shape function. The operator matrix is as follows using the ordering of equations (4.7-4.12).

$$[L] = \begin{bmatrix} \frac{\partial}{\partial r} & 0 & 0 \\ \frac{1}{r} & \frac{1}{r}\frac{\partial}{\partial\vartheta} & 0 \\ \frac{cos\vartheta}{R + rcos\vartheta} & -\frac{sin\vartheta}{R + rcos\vartheta} & \frac{1}{R + rcos\vartheta}\frac{\partial}{\partial\varphi} \\ 0 & \frac{1}{R + rcos\vartheta}\frac{\partial}{\partial\varphi} & \frac{1}{r}\frac{\partial}{\partial\vartheta} + \frac{sin\vartheta}{R + rcos\vartheta} \\ \frac{1}{R + rcos\vartheta}\frac{\partial}{\partial\varphi} & 0 & \frac{\partial}{\partial r} - \frac{cos\vartheta}{R + rcos\vartheta} \\ \frac{1}{r}\frac{\partial}{\partial\vartheta} & \frac{\partial}{\partial r} - \frac{1}{r} & 0 \end{bmatrix} \tag{4.20}$$

The [B] matrix follows from equation (4.19) with the axisymmetric assumptions of equations (4.14) through (4.16) incorporated.

$$[B]=[L][N]=\begin{bmatrix} \frac{\partial N_1}{\partial r} & 0 & 0 & \cdots \\ \frac{N_1}{r} & \frac{1}{r}\frac{\partial N_1}{\partial \vartheta} & 0 & \cdots \\ \frac{N_1 cos\vartheta}{R+rcos\vartheta} & -\frac{N_1 sin\vartheta}{R+rcos\vartheta} & \frac{nN_1}{R+rcos\vartheta} & \cdots \\ 0 & -\frac{nN_1}{R+rcos\vartheta} & \frac{1}{r}\frac{\partial N_1}{\partial \vartheta}+\frac{N_1 sin\vartheta}{R+rcos\vartheta} & \cdots \\ -\frac{nN_1}{R+rcos\vartheta} & 0 & \frac{\partial N_1}{\partial r}-\frac{N_1 cos\vartheta}{R+rcos\vartheta} & \cdots \\ \frac{1}{r}\frac{\partial N_1}{\partial \vartheta} & \frac{\partial N_1}{\partial r}-\frac{N_1}{r} & 0 & \cdots \end{bmatrix} \quad (4.21)$$

where [B] is a 6x3m matrix with m being the number of assumed shape functions. The numerical results that will be discussed were obtained using a nine node isoparametric Lagrangian element. A discussion of Lagrangian finite element shape functions can be found in Buchanan (1995).

The differential volume for axisymmetric toroidal coordinates becomes

$$dV = h_r\vec{e}_r \cdot h_\vartheta\vec{e}_\vartheta \times h_\varphi\vec{e}_\varphi = 2\pi(R + rcos\vartheta)rdrd\vartheta \quad (4.22)$$

Numerical results based upon the theory in this section will be tabulated for several values of R and n. All variables will be made nondimensional in terms outside shell radius a_o, density ρ and shear modulus C_{44}. It follows that a_o=1, ρ=1, C_{44}=1 and the required nondimensional terms are

$$\Omega = \omega a_o\sqrt{\rho/C_{44}}, \quad R = \delta a_o, \qquad b = \lambda a_o, \quad \bar{C}_{kj} = C_{kj}/C_{44}, \quad (4.23)$$

where Ω is the nondimensional frequency, δ is a number greater than 1, b is the inside radius of the shell, λ is a number less than 1 and ν=0.3 is Poisson's ratio.

Very accurate results for frequency were computed by Zhou et al. (2002) using energy methods and results were tabulated for R=1.5 and values of n=0, 1 and 5, as defined in equations (4.14-4.16). Comparison is shown in Table 4.1 where it is shown that the finite element analysis illustrated herein is sufficiently accurate. Values of n=2, 3 and 4 are included in order to make the analysis complete. Several mode shapes were given by Zhou et al. (2002). Table 4.1 is an abridged version of the table that first appeared in Buchanan and Liu (2005). Additional results for various values of R and b were tabulated by Buchanan and Liu (2005) and will not be repeated. However, some additional results will be included in order to make this section more complete.

Results for free vibration were given by Buchanan and Liu (2005) in terms of tables of nondimensional frequency and plots of mode shapes for selected frequencies and the primary focus concerned thick shells. Results for a relatively thin shell will be tabulated in this section. The presentation based upon the elasticity equations will be more complete and it will be possible to compare an elasticity analysis with thin shell analysis and at the same time verify the accuracy of the thin shell analysis given in Section 3.

Table 4.1. Comparison of frequencies $\Omega = \omega a_o\sqrt{\rho/C_{44}}$ with Zhou, Au, Lo and Cheung (2002) for solid toroidal shells with R = 1.5, ν = 0.3

mode	[11]	FE	[11]	FE	FE	FE	FE	[11]	FE
	n = 0	n = 0	n = 1	n = 1	n = 2	n = 3	n = 4	n = 5	n = 5
1	0.8027	0.8028	1.1550	1.1552	0.5247	1.1062	1.6521	2.1631	2.1632
2	1.1768	1.1772	1.1857	1.1858	0.7078	1.4144	1.9801	2.4612	2.4617
3	1.9137	1.9137	2.1434	2.1438	1.6923	2.2781	2.7270	3.1562	3.1573
4	2.2610	2.2611	2.3340	2.3355	1.7830	2.2810	2.7716	3.1897	3.1903
5	2.5206	2.5224	2.4920	2.4934	2.4321	2.7698	3.2356	3.7096	3.7115
6	2.6143	2.6161	2.7232	2.7537	2.4914	2.8840	3.3321	3.7719	3.7745
7	2.8790	2.8797	3.0481	3.0486	2.7454	3.2803	3.8167	4.2982	4.2994
8	2.9885	2.9892	3.1084	3.1089	3.0362	3.4016	3.8187	4.3064	4.3101
9	3.2662	3.2670	3.4698	3.4733	3.3642	3.5964	3.9898	4.4074	4.4129
10	3.2927	3.2935	3.5034	3.5071	3.4993	3.8393	4.2335	4.6849	4.6940
11	3.7970	3.8055	3.7284	3.7316	3.6386	3.9121	4.3690	4.8647	4.8762
12	3.8030	3.8120	3.7716	3.7780	3.7027	4.1462	4.6507	5.1031	5.1087

Table 4.2. Comparison of frequencies $\Omega = \omega a_o\sqrt{\rho/G}$ for toroidal shells with inside radius b=0.9a_o based upon an elasticity analysis and a thin shell analysis. G=C_{44} and ν=0.3

	R=2a_o		R=5a_o		R=10a_o	
n	elasticity	thin shell	elasticity	thin shell	elasticity	thin shell
0	0.1658 (1)	0.1604 (1)	0.1135 (3)	0.1102 (3)	0.0873 (5)	0.0849(5)
	0.4678 (8)	0.4642 (8)	0.2602	0.2560	0.1612	0.1592
	0.6339	0.6228	0.3382	0.3361	0.1922	0.1867
	0.7125	0.6972	0.3464	0.3455	0.2037	0.1988
1	0.4078 (4)	0.4047 (4)	0.1898 (6)	0.1868 (6)	0.1256 (8)	0.1227(8)
	0.4322 (7)	0.4291 (7)	0.2545	0.2603	0.1880	0.1827
	0.6644	0.6497	0.3517	0.3502	0.2060	0.2029
2	0.2579 (2)	0.2573 (2)	0.0698 (1)	0.0689 (1)	0.0279 (1)	0.0214(1)
	0.3056 (3)	0.3038 (3)	0.0715 (2)	0.0698 (2)	0.0320 (2)	0.0237(2)
	0.4117 (5)	0.4106 (5)	0.2347 (7)	0.2316 (7)	0.1730	0.1674
	0.4246 (6)	0.4241 (6)	0.2491 (10)	0.2458 (10)	0.1873	0.1812
3	0.5403 (9)	0.5401 (9)	0.1626 (4)	0.1613 (4)	0.0658 (3)	0.0622(3)
	0.5471 (10)	0.5469 (10)	0.1678 (5)	0.1655 (5)	0.0665 (4)	0.0628(4)
	0.6478	0.6417	0.2818	0.2803	0.1831	0.1787
4	0.7679	0.7663	0.2408 (8)	0.2391 (8)	0.1124 (6)	0.1100(6)
	0.7691	0.7672	0.2434 (9)	0.2410 (9)	0.1136 (7)	0.1110(7)
5	0.9117	0.9105	0.3197	0.3182	0.1564 (9)	0.1537(9)
	0.9124	0.9112	0.3206	0.3189	0.1581 (10)	0.1546(10)

Table 4.2 shows the comparison for shells with thin shell thickness h=0.1a_o from Section 3 versus an elasticity analysis with inside radius designated b=0.9a_o. The results compare nicely and it can be concluded that the analysis of Section 3 is more than satisfactory. Table 4.3 makes a similar comparison for a thin shell with thickness h=0.2a_o. The lowest frequency for R=2 a_o agrees to within 2.5 percent. It can be concluded that thin shell theory that includes shear deformation and rotary inertia can be used for moderately thick shells.

Table 4.3. Comparison of frequencies $\Omega = \omega a_o\sqrt{\rho/G}$ for toroidal shells with inside radius b=0.8 based upon an elasticity analysis and a thin shell analysis. G=C_{44} and v=0.3

	R=$2a_o$		R=$5a_o$		R=$10a_o$	
n	elasticity	thin shell	elasticity	thin shell	elasticity	thin shell
0	0.2699 (1)	0.2631 (1)	0.1815 (3)	0.1790 (3)	0.1078 (5)	0.1072 (5)
	0.5566 (8)	0.5528 (8)	0.3164 (7)	0.3155 (7)	0.1618 (6)	0.1616 (9)
	0.8155	0.8094	0.4081	0.4038	0.3394	0.3330
	0.9636	0.9454	0.4188	0.4150	0.3410	0.3347
1	0.5064 (5)	0.5034 (5)	0.2642 (6)	0.2619 (6)	0.1469 (8)	0.1461 (8)
	0.5342 (7)	0.5307 (7)	0.3856	0.3808	0.2262	0.2262
	0.7671	0.7630	0.4143	0.4110	0.3392	0.3327
2	0.3278 (2)	0.3290 (2)	0.0839 (1)	0.0842 (1)	0.0246 (1)	0.0256 (1)
	0.4038 (3)	0.3990 (3)	0.0877 (2)	0.0873 (2)	0.0257 (2)	0.0261 (2)
	0.4886 (4)	0.4902 (4)	0.3631 (9)	0.3593 (9)	0.2273	0.2267
	0.5157 (6)	0.5184 (6)	0.3777	0.3739	0.3384	0.3326
3	0.6777 (9)	0.6811 (9)	0.2118 (4)	0.2119 (4)	0.0683 (3)	0.0684 (3)
	0.6966 (10)	0.7010 (10)	0.2227 (5)	0.2217 (5)	0.0699 (4)	0.0698 (4)
	0.7537	0.7533	0.3858	0.3843	0.3147	0.3126
4	0.9136	0.9173	0.3545 (8)	0.3539 (8)	0.1236 (6)	0.1238 (6)
	0.9248	0.9292	0.3682 (10)	0.3669 (10)	0.1262 (7)	0.1263 (7)
5	1.1283	1.1327	0.4639	0.4654	0.1882 (10)	0.1884 (10)
	1.1349	1.1399	0.4664	0.4681	0.1916	0.1914

Table 4.4. Frequencies $\Omega = \omega a_o\sqrt{\rho/G}$ for toroidal shells with 80 degree opening, fixed boundary conditions and v=0.3

	b=$0.8a_o$			b=$0.9a_o$		
n	R=$2a_o$	R=$5a_o$	R=$10a_o$	R=$2a_o$	R=$5a_o$	R=$10a_o$
0	0.1564 (1)	0.1443 (1)	0.1443 (1)	0.0959 (1)	0.0757 (1)	0.0718 (1)
	0.3934 (3)	0.4280 (5)	0.3863 (8)	0.3582 (4)	0.2830 (4)	0.2145 (5)
	0.5890 (7)	0.5721	0.6410	0.4767 (7)	0.4181	0.3703
	0.8754	0.7432	0.7181	0.6058 (10)	0.5344	0.5728
1	0.2043 (2)	0.1681 (2)	0.1529 (2)	0.1656 (2)	0.1128 (2)	0.0889 (2)
	0.4446 (5)	0.4290 (6)	0.3880 (9)	0.3511 (3)	0.2847 (5)	0.2183 (6)
	0.7916 (10)	0.6390	0.6578	0.5717 (8)	0.4201	0.3723
2	0.4372 (4)	0.2356 (3)	0.1776 (3)	0.4162 (5)	0.1983 (3)	0.1225 (3)
	0.5617 (6)	0.4435 (7)	0.3935 (10)	0.4624 (6)	0.3036 (6)	0.2244 (8)
	0.8467	0.7530	0.7057	0.5807 (9)	0.4308	0.3757
	1.0989	0.7988	0.7226	0.7165	0.6087	0.5744
3	0.7307 (8)	0.3355 (4)	0.2145 (4)	0.6386	0.3053 (7)	0.1698 (4)
	0.7431 (9)	0.4768 (9)	0.4029	0.6583	0.3436 (8)	0.2388 (9)
	0.9690	0.7716	0.7285	0.7350	0.4555	0.3846
4	0.9453	0.4527 (8)	0.2609 (5)	0.7894	0.4019 (9)	0.2217 (7)
	0.9634	0.5283 (10)	0.4165	0.7951	0.4085 (10)	0.2578 (10)
5	1.1418	0.5735	0.3137 (6)	0.9230	0.4704	0.2735
	1.1532	0.5951	0.4348	0.9265	0.4794	0.2825
6	1.3289	0.6733	0.3711 (7)	1.0528	0.5377	0.3119
	1.3344	0.6869	0.4577	1.0535	0.5430	0.3223

Incomplete axisymmetric toroidal shells can be studied using the methodology of this section. Incomplete, in the sense, that the toroidal shell is a sector of a complete circular torus.

Two shells will be discussed and are shown in Figure 4.1. One has a symmetrical opening in the interior of the shell of 80 degrees. The second is a half shell with an opening of 180 degrees. Both shells are completely fixed at their ends.

That is, U=V=W=0 at the ends of the shell sector. Frequency results are tabulated in Tables 4.4 and 4.5 for shell inside radius of $0.8a_o$ and $0.9a_o$. The tables show the frequency vibration behavior of the shells and in every case sufficient frequencies are included to show the distribution of the first ten frequencies. Table 4.5 includes results up to and including n=9 in order to present the first ten frequencies.

Table 4.5. Frequencies $\Omega = \omega a_o\sqrt{\rho/G}$ for toroidal shells with 180 degree opening, fixed boundary conditions and ν=0.3

	b=$0.8a_o$			b=$0.9a_o$		
n	R=$2a_o$	R=$5a_o$	R=$10a_o$	R=$2a_o$	R=$5a_o$	R=$10a_o$
0	0.4844 (1)	0.4797 (1)	0.4819 (1)	0.2852 (1)	0.2481 (1)	0.2375 (1)
	0.8677 (5)	0.9787	0.9648	0.6502 (4)	0.5416 (7)	0.5136
	1.0443 (8)	0.9988	1.0527	0.8128 (9)	0.9403	0.9281
	1.7357	1.7089	1.7072	1.0031	0.9404	0.9933
1	0.5282 (2)	0.4890 (2)	0.4845 (3)	0.3508 (2)	0.2658 (2)	0.2425 (2)
	0.9513	0.9782 (10)	0.9656	0.6541 (5)	0.5460 (8)	0.5140
	1.1119 (9)	1.0356	1.0622	0.9572	0.9433	0.9289
2	0.6533 (3)	0.5171 (3)	0.4921 (2)	0.5082 (3)	0.3083 (3)	0.2564 (3)
	1.0346 (6)	0.9852	0.9677	0.6970 (6)	0.5556 (9)	0.5179
	1.3619	1.1301	1.0904	1.0381	0.9499	0.9314
	1.7297	1.7128	1.7090	1.2070	1.0668	1.0323
3	0.8349 (4)	0.5630 (4)	0.5049 (4)	0.6990 (7)	0.3730 (4)	0.2771 (4)
	1.1138 (10)	0.9990	0.9714	0.7747 (8)	0.5737 (10)	0.5216
	1.5981	1.2638	1.1353	1.0964	0.9626	0.9346
4	1.0376 (7)	0.6250 (5)	0.5227 (5)	0.8753 (10)	0.4508 (5)	0.3050 (5)
	1.2130	1.0193	0.9766	0.8766	0.5993	0.5280
5	1.2343	0.7000 (6)	0.5453 (6)	0.9869	0.5354 (6)	0.3381 (6)
	1.3286	1.0464	0.9834	1.0173	0.6319	0.5360
6	1.4156	0.7844 (7)	0.5728 (7)	1.1029	0.6210	0.3754 (7)
	1.4524	1.0801	0.9919	1.1332	0.6710	0.5467
7	1.5777	0.8745 (8)	0.6045 (8)	1.2185	0.7042	0.4152 (8)
	1.5864	1.1205	1.0022	1.2400	0.7161	0.5591
8	1.7031	0.9672 (9)	0.6403 (9)	1.3321	0.7650	0.4577 (9)
	1.7556	1.1669	1.0144	1.3443	0.7812	0.5743
9	1.8320	1.0599	0.6794(10)	1.4432	0.8186	0.5014 (10)
	1.9305	1.2189	1.0284	1.4483	0.8521	0.5907

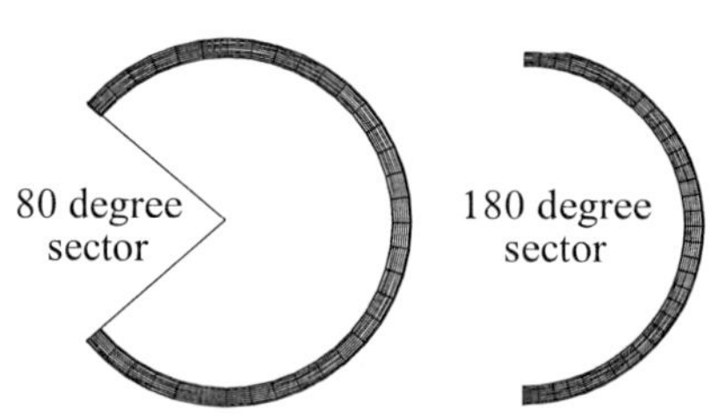

Figure 4.1. Toroidal shell as a sector of a circular shell.

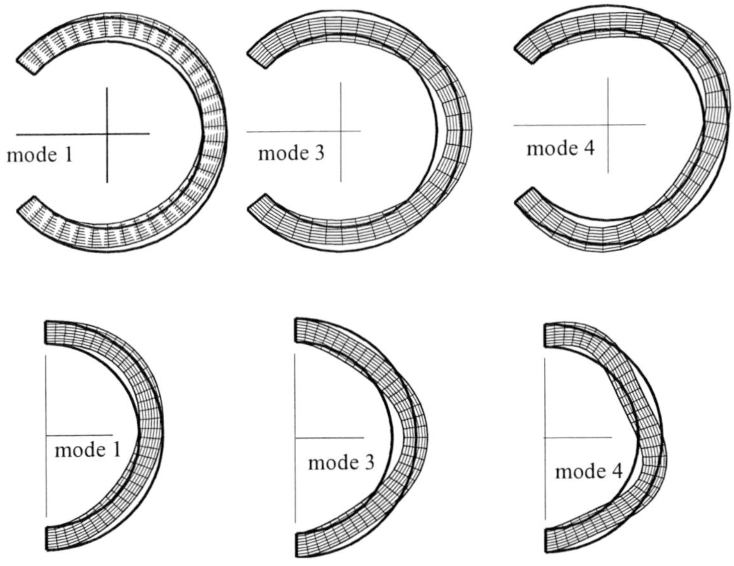

Figure 4.2. Modes shapes for sectors of toroidal shells shown in Figure 4.1 with h=0.8a and n=0.

Mode shapes for the case n=0, R=2 and inside radius b=0.8 are shown in Figure 4.2. In both cases the second mode is torsional and is not shown. Figure 4.2 is representative of information that can be obtained using the concepts presented in this section.

5. Three-Dimensional Elasticity Equations and Analysis in a Cylindrical Coordinate System for Noncircular Toroidal Shell Cross-Sections

Toroidal shells with noncircular cross-section can be studied using a cylindrical coordinate system (r,ϑ,z) with suitable assumptions that render the system axisymmetric. Any cross-section that can be modeled in an axisymmetric (r,z) coordinate space can represent the cross-section of a toroidal shell. In this section an elliptic cross-section will be studied and a vertical oblong section with spherical end caps will be proposed. The intent is that the reader will realize the significance and versatility when using axisymmetric cylindrical coordinate system to model a toroidal shell.

The three-dimensional strain-displacement equations of elasticity in cylindrical coordinates (r,ϑ,z) are the initial necessary equations for developing an axisymmetric finite element in the (r,z) space.

Any elasticity text, such as Chow and Pagano (1967), can be consulted to obtain the strain-displacement equations. In the current section the displacements will be denoted u, v, and w to correspond with cylindrical coordinates (r,θ,z). The matrix ordering that was used in Section 4 will be used in this section.

$$\varepsilon_{rr} = \varepsilon_1 = \frac{\partial u}{\partial r} \qquad (5.1)$$

$$\varepsilon_{\vartheta\vartheta} = \varepsilon_2 = \frac{u}{r} + \frac{1}{r}\frac{\partial v}{\partial \vartheta} \tag{5.2}$$

$$\varepsilon_{zz} = \varepsilon_3 = \frac{\partial w}{\partial z} \tag{5.3}$$

$$\varepsilon_{\vartheta z} = \varepsilon_4 = \frac{\partial v}{\partial z} + \frac{1}{r}\frac{\partial w}{\partial \vartheta} \tag{5.4}$$

$$\varepsilon_{rz} = \varepsilon_5 = \frac{\partial w}{\partial r} + \frac{\partial u}{\partial z} \tag{5.5}$$

$$\varepsilon_{r\vartheta} = \varepsilon_6 = \frac{\partial v}{\partial r} - \frac{v}{r} + \frac{1}{r}\frac{\partial u}{\partial \vartheta} \tag{5.6}$$

Again, as in the previous sections, periodic type solutions can be shown to satisfy the displacement equations of motion and subsequently reduce the finite element formulation from three to two dimensions while preserving a three-dimensional analysis. Accordingly, assume,

$$u(r,\vartheta,z,t) = U(r,z)cosn\vartheta cos\omega t \tag{5.7}$$

$$v(r,\vartheta,z,t) = V(r,z)sinn\vartheta cos\omega t \tag{5.8}$$

$$w(r,\vartheta,z,t) = W(r,z)cosn\vartheta cos\omega t \tag{5.9}$$

where n is the circumferential wave number and ω is the circular frequency.

Free vibration analysis is defined as in the previous Sections in terms of the stiffness matrix and mass matrix.

$$([K] - \omega^2[M])\{d\} = 0 \tag{5.10}$$

The stiffness matrix [K] is defined as,

$$[K] = \int_V [B]^T\,[C]\,[B]dV \tag{5.11}$$

The material matrix [C] remains the same as equation (4.13) and the [B] matrix is,

$$[B] = [L]\,[N] \tag{5.12}$$

The operator matrix [L] is obtained in terms of the strain-displacement equations (5.1) through (5.6) and [N] is a suitable two-dimensional shape function. The operator matrix is as follows,

$$[L] = \begin{bmatrix} \frac{\partial}{\partial r} & 0 & 0 \\ \frac{1}{r} & \frac{1}{r}\frac{\partial}{\partial \vartheta} & 0 \\ 0 & 0 & \frac{\partial}{\partial z} \\ 0 & \frac{\partial}{\partial z} & \frac{1}{r}\frac{\partial}{\partial \vartheta} \\ \frac{\partial}{\partial z} & 0 & \frac{\partial}{\partial r} \\ \frac{1}{r}\frac{\partial}{\partial \vartheta} & \frac{\partial}{\partial r} - \frac{1}{r} & 0 \end{bmatrix} \tag{5.13}$$

The [B] matrix, incorporating equations (5.7-5.9) becomes

$$[B] = \begin{bmatrix} \frac{\partial N_1}{\partial r} & 0 & 0 & \cdots\cdots \\ \frac{N_1}{r} & \frac{nN_1}{r} & 0 & \cdots\cdots \\ 0 & 0 & \frac{\partial N_1}{\partial z} & \cdots\cdots \\ 0 & \frac{\partial N_1}{\partial z} & -\frac{nN_1}{r} & \cdots\cdots \\ \frac{\partial N_1}{\partial z} & 0 & \frac{\partial N_1}{\partial r} & \cdots\cdots \\ -\frac{nN_1}{r} & \frac{\partial N_1}{\partial r} - \frac{N_1}{r} & 0 & \cdots\cdots \end{bmatrix} \tag{5.14}$$

There are additional columns for the remaining shape functions. The numerical results that will be discussed were obtained using the same nine node isoparametric Lagrangian element as in the previous section.

5.1. Elliptical Cross Section

Vibration of a toroidal shell with elliptical cross-section has been studied by Yamada et al. (1989) based upon thin shell theory. Later, Xu and Redekop (2006) extended the study to include orthotropic material properties. In both publications the analysis was based upon thin shell theory neglecting shear deformation and rotary inertia.

The cross-section of the elliptical torus in an axisymmetrical (r,z) coordinate system is shown in Figure 5.1 and can be either solid or hollow. The finite element mesh was generated using the equation for an ellipse as,

$$\frac{x^2}{a^2} + \frac{y^2}{b^2} = 1 \tag{5.15}$$

where a and b are the major and minor dimensions of the ellipse, respectively. The equations for an ellipse in elliptical coordinates can be used but they offer the disadvantage of creating a finite element mesh of conformal ellipses making it more difficult to control a constant thickness for a thick-walled hollow cross-section. The author and co-workers (Liu et al. (2006), Madhavapeddy (2006)) have some experience with the conformal type of formulation. It follows that thin shell equations could be formulated in elliptical coordinates and the methods of Sections 2 and 3 used to study the vibration of thin elliptical toroidal shells.

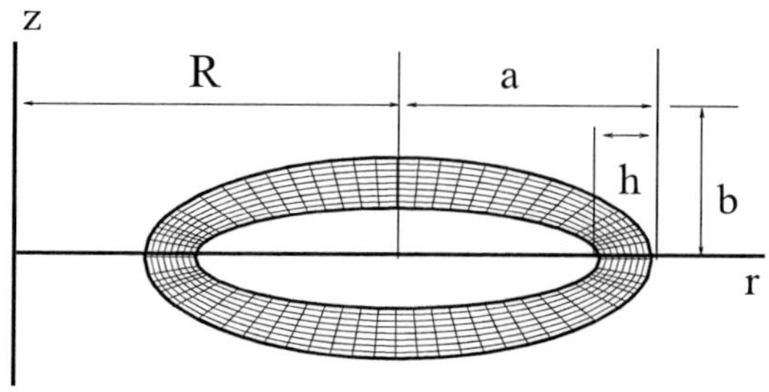

Figure 5.1. Elliptic toroidal shell in an axisymmetric cylindrical (r,z) coordinate system.

Again, limited results are given to illustrate the analysis and to provide some benchmark solutions that might prove useful to future researchers. The frequency is rendered nondimensional in terms of minor axes b, whereby b=1., ρ=1. and G=1. Nondimensional frequency is given as,

$$\Omega = \omega b \sqrt{\rho / G} \tag{5.16}$$

The accuracy and versatility of the formulation is demonstrated in Table 5.1 where the major axis is assumed as a=1.01 in order to approximate a circular cross section. The results compare favorably with the frequencies given in Table 4.3 of the previous section.

Table 5.1. Comparison of frequencies for circular torus using the elasticity analysis of Section 4, Table 4.3, and an axisymmetric elliptical cross-section with a=1.01

mode	R=2.0		R=5.0	
	elasticity	axisymmetric	elasticity	axisymmetric
0	0.2699	0.2689	0.1815	0.1813
1	0.5064	0.5055	0.2642	0.2640

Additional results are given in Table 5.2 for shell thickness of h=.2b and h=.1b. Sufficient results are tabulated to show the behavior of the first ten frequencies. The first ten frequencies occur within the first three modes of vibration. Typical mode shapes for the first three modes when n=0 are shown in Figure 5.2.

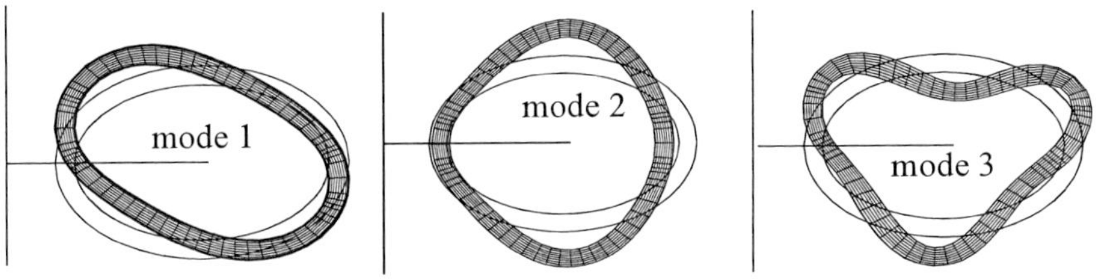

Figure 5.2. Modes shapes for elliptical cross-section, R=2b, a=1.5b, h=.2b.

Table 5.2. Frequencies $\Omega = \omega b\sqrt{\rho/G}$ for elliptical toroidal shells with b=1.0, a=1.5b, a=2.5b, h=0.1b, h=0.2b and ν=0.3

	a=1.5b				a=2.5b	
mode	R=2.0b		R=5.0b		R=5.0b	
	h=0.1b	h=0.2b	h=0.1b	h=0.2b	h=0.1b	h=0.2b
0	0.1004 (1)	0.1648 (1)	0.0636 (1)	0.1110 (2)	0.0315 (1)	0.0515 (1)
	0.2944 (5)	0.3596 (4)	0.1730 (9)	0.2344 (8)	0.0853 (5)	0.1100 (5)
	0.4232	0.6234	0.2846	0.3471	0.1399	0.2251
	0.4577	0.6785	0.3037	0.3588	0.1442	0.2345
1	0.2399 (2)	0.3044 (2)	0.1336 (4)	0.1824 (4)	0.0788 (4)	0.1032 (4)
	0.3251 (6)	0.3878 (6)	0.1667 (8)	0.2362 (9)	0.0914 (7)	0.1119 (6)
	0.3928 (10)	0.5632	0.2918	0.3659	0.1364	0.2237
	0.3986	0.6222	0.3012	0.4219	0.1381	0.2320
	0.6112	0.7456	0.3415	0.5326	0.2305	0.3103
2	0.2849 (3)	0.3045 (3)	0.0678 (2)	0.0825 (1)	0.0641 (2)	0.0782 (2)
	0.2878 (4)	0.3661 (5)	0.0879 (3)	0.1197 (3)	0.0762 (3)	0.0979 (3)
	0.3447 (7)	0.4393 (7)	0.1575 (7)	0.2251 (6)	0.0872 (6)	0.1233 (7)
	0.3569 (8)	0.5227 (9)	0.1757 (10)	0.2313 (7)	0.1272 (10)	0.1875 (10)
	0.3715 (9)	0.6052	0.2865	0.4689	0.1299	0.2160
	0.5348	0.7181	0.2964	0.5388	0.1912	0.2348
3	0.4471	0.5077 (8)	0.1512 (5)	0.2090 (5)	0.1100 (8)	0.1398 (8)
	0.4628	0.5571 (10)	0.1556 (6)	0.2366 (10)	0.1146 (9)	0.1426 (9)
	0.4677	0.6729	0.2183	0.2520	0.1368	0.2004
4	0.5612	0.6764	0.2174	0.2973	0.1506	0.1893
	0.5803	0.7123	0.2266	0.3055	0.1545	0.1956
5	0.6636	0.8280	0.2961	0.3716	0.1840	0.2330
	0.6831	0.8633	0.3028	0.3797	0.1855	0.2455
6	0.7590	0.9733	0.3657	0.4527	0.2145	0.2740
	0.7803	1.0195	0.3765	0.4632	0.2169	0.2892

5.2. Oblong Cross-Section with Spherical Caps

A final example is based upon a completely arbitrary cross-section that is shown in Figure 5.3 and is referred to as an oblong section with spherical end caps. The dimensions are nondimensional with respect to the radius of the spherical end cap that is equal 1. The toroidal radius is R, the distance between the end caps is L and the thickness is h. It follows that nondimensional frequency is given as

$$\Omega = \omega a\sqrt{\rho/G} \tag{5.17}$$

Table 5.3. Frequencies $\Omega = \omega b\sqrt{\rho/G}$ for toroidal shell, oblong with spherical caps with a=1.0, L=2.0a, h=0.1a and 0.2a for R=2.5a and R=5.0a

	R=2.5a		R=5.0a	
mode	h=0.1a	h=0.2a	h=0.1a	h=0.2a
0	0.1080 (1)	0.1649 (1)	0.0756 (3)	0.1211 (3)
	0.4231	0.4615	0.2622	0.2823
	0.4555	0.5274	0.2857	0.3222
1	0.3382	0.3678 (7)	0.1862 (8)	0.2169 (7)
	0.3396	0.3750 (9)	0.2539	0.2870
	0.4564	0.5645	0.2965	0.3144
2	0.1494 (2)	0.2013 (2)	0.04587 (1)	0.0634 (1)
	0.1670 (3)	0.2226 (3)	0.05750 (2)	0.0798 (2)
	0.2625 (7)	0.2871 (4)	0.2054	0.2382 (9)
	0.3456	0.4165 (11)	0.2329	0.2641
	0.4032	0.4995	0.3085	0.4012
3	0.2601 (6)	0.3152 (5)	0.1014 (4)	0.1446 (4)
	0.2796 (9)	0.4145 (10)	0.1311 (5)	0.1817 (5)
	0.3454	0.5165	0.2087	0.2362 (8)
	0.4856	0.5391	0.2751	0.3495
4	0.2592 (5)	0.3355 (6)	0.1371 (6)	0.2023 (6)
	0.3920	0.6420	0.1968 (9)	0.2595
5	0.2538 (4)	0.3748 (8)	0.1663 (7)	0.2499 (10)
	0.5219	0.7216	0.2048	0.2866
6	0.2695 (8)	0.4397 (12)	0.1991 (10)	0.2824
	0.5440	0.7982	0.2052	0.3352
7	0.3019 (10)	0.5278	0.2087	0.3087
	0.5739	0.8832	0.2368	0.4044

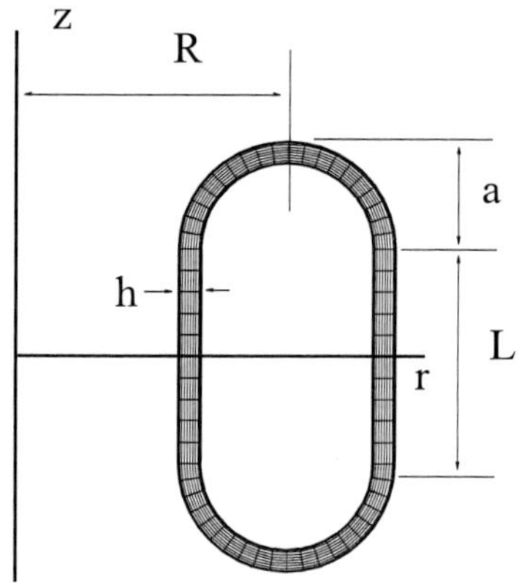

Figure 5.3. Oblong with spherical caps toroidal shell in an axisymmetric cylindrical (r,z) coordinate system.

An example was assumed for the oblong cross-section with L=2a and some typical values for R and h.

The results for nondimensional frequency are given in Table 5.3. The first ten frequencies are identified in order to show the effect of the circular parameter n. Representative mode shapes are shown in Figure 5.4 for the case n=0, R=2.5a, L=2a and h=0.2a.

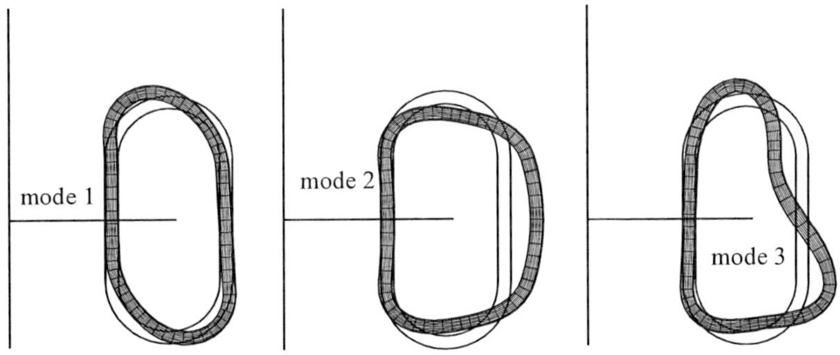

Figure 5.4. Modes shapes for oblong cross-section, free boundary conditions, a=1., R=2.5a, L=2a, h=.2a, n=0.

6. Free Vibration of Toroidal Shells with Transversely Isotropic Material Properties

An extension of fundamental solution techniques to include transversely isotropic material properties is a logical next topic and a fitting conclusion for the basic concepts that have been developed and discussed thus far. The crystal physics counterpart of the engineering definition of transversely isotropic would be the crystal classification hexagonal. Nye (1969) can be consulted for a discussion of the classical topic of crystal physics. The concept of transversely isotropic materials is embedded in the vast literature on composite materials. There are numerous textbooks that give an excellent discussion of composite materials, for instance, Hyer (1997) or Jones (1999). A general discussion of transversely isotropic materials from a mechanics viewpoint can be found in the text by Bisplinghoff, Mar, and Pian (1965). The ideas and concepts presented here are an extension of work published by Liu and Buchanan (2006).

6.1. Three-Dimensional Elasticity Analysis

Begin the discussion with reference to equation (4.13) where the isotropic constitutive equations were written to correspond to a toroidal coordinate system. Constitutive equations are independent of the coordinate system and equation (4.13) was also used in Section 5 for cylindrical coordinates. The same invariance holds for transversely isotropic material constants, however, some care must be exercised when establishing the axis about which the transverse isotropy occurs. A material can be termed transversely isotropic with respect to a given axis and in Cartesian coordinates equation (6.1) would be transversely isotropic with respect to the z axis. That is, the material constants remain unchanged with respect to coordinate rotation about the z axis. Sometimes, the terminology is referred to as poled with respect to the z axis.

Again, the order in which σ_4, σ_5, and σ_6 are defined is arbitrary and an order different than that used in equation (6.1) will change the location of shear material constants in the material matrix.

$$\begin{Bmatrix} \sigma_{xx} \\ \sigma_{yy} \\ \sigma_{zz} \\ \sigma_{yz} \\ \sigma_{xz} \\ \sigma_{xy} \end{Bmatrix} = \begin{Bmatrix} \sigma_1 \\ \sigma_2 \\ \sigma_3 \\ \sigma_4 \\ \sigma_5 \\ \sigma_6 \end{Bmatrix} = [C]\{\varepsilon\} = \begin{bmatrix} C_{11} & C_{12} & C_{13} & 0 & 0 & 0 \\ C_{12} & C_{11} & C_{13} & 0 & 0 & 0 \\ C_{13} & C_{13} & C_{33} & 0 & 0 & 0 \\ 0 & 0 & 0 & C_{44} & 0 & 0 \\ 0 & 0 & 0 & 0 & C_{44} & 0 \\ 0 & 0 & 0 & 0 & 0 & C_{66} \end{bmatrix} \begin{Bmatrix} \varepsilon_1 \\ \varepsilon_2 \\ \varepsilon_3 \\ \varepsilon_4 \\ \varepsilon_5 \\ \varepsilon_6 \end{Bmatrix} \tag{6.1}$$

For materials transversely isotropic relative to the z axis it follows that

$$C_{66} = (C_{11} - C_{12})/2. \tag{6.2}$$

Let the material be poled relative to the x axis, but keep the same order for the stresses as in equation (6.1). The material matrix is given as,

$$\begin{Bmatrix} \sigma_{xx} \\ \sigma_{yy} \\ \sigma_{zz} \\ \sigma_{yz} \\ \sigma_{xz} \\ \sigma_{xy} \end{Bmatrix} = \begin{Bmatrix} \sigma_1 \\ \sigma_2 \\ \sigma_3 \\ \sigma_4 \\ \sigma_5 \\ \sigma_6 \end{Bmatrix} = \begin{bmatrix} C_{33} & C_{13} & C_{13} & 0 & 0 & 0 \\ C_{13} & C_{11} & C_{12} & 0 & 0 & 0 \\ C_{13} & C_{12} & C_{11} & 0 & 0 & 0 \\ 0 & 0 & 0 & C_{66} & 0 & 0 \\ 0 & 0 & 0 & 0 & C_{44} & 0 \\ 0 & 0 & 0 & 0 & 0 & C_{44} \end{bmatrix} \begin{Bmatrix} \varepsilon_1 \\ \varepsilon_2 \\ \varepsilon_3 \\ \varepsilon_4 \\ \varepsilon_5 \\ \varepsilon_6 \end{Bmatrix} \tag{6.3}$$

Equation (6.2) remains valid for C_{66}.

Table 6.1. Elastic constants for cobalt and zinc from Payton (1983). The notation corresponds to equation (6.1). Multiply by 10^9 N/m^2

material	C_{11}	C_{33}	C_{12}	C_{13}	C_{44}	C_{66}
cobalt	307	356	165	103	75.5	71
nondimensional	4.07	4.72	2.19	1.36	1.00	0.94
zinc	165	62	31	50	39.6	67
nondimensional	4.17	1.57	0.78	1.26	1.00	1.69
isotropic, ν=0.3 nondimensional	3.5	3.5	1.5	1.5	1.00	1.00

The examples in this Section will be based upon material constants in the form of stiffness coefficients, such as equations (6.1) and (6.3) because they are necessary in terms of the structure of the finite element equations. Material properties are often recorded in the literature in terms of engineering constants because they are usually determined experimentally in an engineering constant format. Material constants for a transversely isotropic material in terms engineering constants have been recorded by Jiang and Redekop (2002). Material constants that relate strain to stress are termed compliance constants and can be written directly in terms of engineering constants and are usually denoted as S_{kj} or [S]. An example of the 6x6 compliance matrix for orthotropic materials has been given by Jones (1999) and can easily be reduced to transversely isotropic. It follows that the stiffness matrix [C] is the inverse of the compliance matrix [S] and engineering constants can easily be converted to stiffness matrix notation.

Material properties for several materials with hexagonal crystal structure have been recorded by Payton (1983). Cobalt and zinc will be used to illustrate the analysis of transversely isotropic materials. The choice of materials is arbitrary, but when the material constants are made nondimensional they are very different form the nondimensional isotropic material constants when ν=0.3. The materials are contrasted in Table 6.1.

Let the x axis of equation (6.3) correspond to the r axis of Figure 3.1 and it follows that the material constants are poled relative to the r axis. The order of the stresses remains the same as equation (4.13) and the material constants are written as,

$$\begin{Bmatrix} \sigma_{rr} \\ \sigma_{\vartheta\vartheta} \\ \sigma_{\varphi\varphi} \\ \sigma_{\theta\varphi} \\ \sigma_{r\varphi} \\ \sigma_{r\vartheta} \end{Bmatrix} = \begin{Bmatrix} \sigma_1 \\ \sigma_2 \\ \sigma_3 \\ \sigma_4 \\ \sigma_5 \\ \sigma_6 \end{Bmatrix} = [C]\{\varepsilon\} = \begin{bmatrix} C_{33} & C_{13} & C_{13} & 0 & 0 & 0 \\ C_{13} & C_{11} & C_{12} & 0 & 0 & 0 \\ C_{13} & C_{12} & C_{11} & 0 & 0 & 0 \\ 0 & 0 & 0 & C_{66} & 0 & 0 \\ 0 & 0 & 0 & 0 & C_{44} & 0 \\ 0 & 0 & 0 & 0 & 0 & C_{44} \end{bmatrix} \begin{Bmatrix} \varepsilon_1 \\ \varepsilon_2 \\ \varepsilon_3 \\ \varepsilon_4 \\ \varepsilon_5 \\ \varepsilon_6 \end{Bmatrix} \tag{6.4}$$

with $C_{66} = (C_{11} - C_{12})/2$.

The analysis process for the elasticity formulation follows that of Section 4, equation (4.17) with the solution of the standard eigenvalue problem,

$$([K] - \omega^2[M])\{d\} = 0 \tag{6.5}$$

The mass matrix is the same as that used in Section 4.

As in Section 4 all variables will be made nondimensional in terms outside shell radius a_o, density ρ and shear modulus C_{44}. It follows that a_o=1, ρ=1, C_{44}=1 and the required nondimensional terms are the same as Section 4,

$$\Omega = \omega a_o\sqrt{\rho/C_{44}}, \quad R = \delta a_o, b = \lambda a_o, \quad \bar{C}_{kj} = C_{kj}/C_{44}, \tag{6.6}$$

where Ω is the nondimensional frequency, δ is a number greater than 1, b is the inside radius of the shell, λ is a number less than 1. The nondimensional material matrix for cobalt, poled relative to the r axis, $[C_r]$ corresponding to equation (6.4) becomes,

$$[C_r]_{cobalt} = \begin{bmatrix} 4.72 & 1.36 & 1.36 & 0 & 0 & 0 \\ 1.36 & 4.07 & 2.19 & 0 & 0 & 0 \\ 1.36 & 2.19 & 4.07 & 0 & 0 & 0 \\ 0 & 0 & 0 & 0.94 & 0 & 0 \\ 0 & 0 & 0 & 0 & 1.00 & 0 \\ 0 & 0 & 0 & 0 & 0 & 1.00 \end{bmatrix} \tag{6.7}$$

Return to equation (6.1) and assume that the transversely isotropic material is isotropic in the r,θ plane and the material matrix must reflect a material that is poled relative to the φ axis. Maintain the ordering given by the left-hand-side of equation (6.4) and the stress-strain relation becomes.

$$\begin{Bmatrix} \sigma_{rr} \\ \sigma_{\vartheta\vartheta} \\ \sigma_{\varphi\varphi} \\ \sigma_{\theta\varphi} \\ \sigma_{r\varphi} \\ \sigma_{r\vartheta} \end{Bmatrix} = \begin{Bmatrix} \sigma_1 \\ \sigma_2 \\ \sigma_3 \\ \sigma_4 \\ \sigma_5 \\ \sigma_6 \end{Bmatrix} = [C]\{\varepsilon\} = \begin{bmatrix} C_{11} & C_{12} & C_{13} & 0 & 0 & 0 \\ C_{12} & C_{11} & C_{13} & 0 & 0 & 0 \\ C_{13} & C_{13} & C_{33} & 0 & 0 & 0 \\ 0 & 0 & 0 & C_{44} & 0 & 0 \\ 0 & 0 & 0 & 0 & C_{44} & 0 \\ 0 & 0 & 0 & 0 & 0 & C_{66} \end{bmatrix} \begin{Bmatrix} \varepsilon_1 \\ \varepsilon_2 \\ \varepsilon_3 \\ \varepsilon_4 \\ \varepsilon_5 \\ \varepsilon_6 \end{Bmatrix} \tag{6.8}$$

The corresponding nondimensional material matrix $[C_\varphi]$ for cobalt becomes,

$$[C_\varphi]_{cobalt} = \begin{bmatrix} 4.07 & 2.19 & 1.36 & 0 & 0 & 0 \\ 2.19 & 4.07 & 1.36 & 0 & 0 & 0 \\ 1.36 & 1.36 & 4.72 & 0 & 0 & 0 \\ 0 & 0 & 0 & 1.00 & 0 & 0 \\ 0 & 0 & 0 & 0 & 1.00 & 0 \\ 0 & 0 & 0 & 0 & 0 & 0.94 \end{bmatrix} \tag{6.9}$$

6.2. Thin Shell Analysis

The elasticity constitutive equations must be modified in order to use the matrix of material constants as given in equation (2.20). The isotropic constants of equation (2.12) will be replaced with the corresponding transversely isotropic material constants. The concept of reduced stiffness material constants will be discussed in order to make the Section presentation complete. The use of reduced stiffness is common in plate theory and is based upon the plane stress assumption, that is, the stress normal to the surface of the plate or shell is zero.

Assume the case when the material is transversely isotropic relative to the r coordinate axis and equation (6.4) is applicable for the material constants. The plane stress assumption means that σ_{rr} equals zero and the first stress-strain equation can be solved for ε_{rr} and that is substituted into the remaining stress-strain equations. The equations are now written in terms of reduced stiffness material constants.

$$\sigma_{\vartheta\vartheta} = Q_{11}\epsilon_{\vartheta\vartheta} + Q_{12}\epsilon_{\varphi\varphi} \tag{6.10}$$

$$\sigma_{\varphi\varphi} = Q_{12}\epsilon_{\vartheta\vartheta} + Q_{22}\epsilon_{\varphi\varphi} \tag{6.11}$$

$$\sigma_{\vartheta\varphi} = Q_{33}\epsilon_{\vartheta\varphi} \tag{6.12}$$

$$\sigma_{r\varphi} = Q_{44}\epsilon_{r\varphi} \tag{6.13}$$

$$\sigma_{r\vartheta} = Q_{55}\epsilon_{r\vartheta} \tag{6.14}$$

The reduced stiffness material constants are,

$$Q_{11} = C_{11} - \frac{C_{13}^2}{C_{33}}, \qquad Q_{22} = C_{11} - \frac{C_{13}^2}{C_{33}}, Q_{12} = C_{12} - \frac{C_{13}^2}{C_{33}},$$

$$Q_{33} = C_{66}, \ Q_{44} = C_{44}, \qquad Q_{55} = C_{44} \tag{6.15}$$

Transversely isotropic material constants corresponding to the isotropic constants of equation (2.12) become,

$$k_{11} = Q_{11}h\,, \qquad k_{22} = Q_{22}h, \qquad k_{12} = Q_{12}h, k_{33} = Q_{33}h$$

$$D_{11} = \frac{Q_{11}h^3}{12}, D_{22} = \frac{Q_{22}h^3}{12}, D_{12} = \frac{Q_{12}h^3}{12}, D_{33} = \frac{Q_{33}h^3}{12} \tag{6.16}$$

$$G_{13} = G_{23} = k'C_{44}h$$

The reduced stiffness material constants corresponding to equation (6.8) for the material poled relative to the φ axis are derived using the same process, that is, $\sigma_{rr}=0$ and because the elements of the material matrix are modified the corresponding reduced stiffness will also be different from equation (6.15). The reduced stiffness for material poled relative to the φ axis become,

$$Q_{11} = C_{11} - \frac{C_{12}^2}{C_{11}}, \qquad Q_{22} = C_{33} - \frac{C_{13}^2}{C_{11}}, Q_{12} = C_{13} - \frac{C_{12}C_{13}}{C_{11}},$$

$$Q_{33} = C_{66}, \ Q_{44} = C_{44}, \qquad Q_{55} = C_{44} \tag{6.17}$$

The subscripts for the Q values in equation (6.17) correspond to equation (6.16) and the final material constants are computed as in equation (6.16).

The nondimensional material matrix $[\overline{D}]$ as defined in equations (2.17) and (2.20) for cobalt poled relative to the toroidal r axis with shell thickness h=0.1a_o becomes,

$$[\overline{D}_r] = \begin{bmatrix} 0.3678 & 0.1798 & 0 & 0 & 0 & 0 & 0 & 0 \\ 0.1798 & 0.3678 & 0 & 0 & 0 & 0 & 0 & 0 \\ 0 & 0 & 0.094 & 0 & 0 & 0 & 0 & 0 \\ 0 & 0 & 0 & 30.65(10^{-5}) & 14.98(10^{-5}) & 0 & 0 & 0 \\ 0 & 0 & 0 & 14.98(10^{-5}) & 30.65(10^{-5}) & 0 & 0 & 0 \\ 0 & 0 & 0 & 0 & 0 & 7.83(10^{-5}) & 0 & 0 \\ 0 & 0 & 0 & 0 & 0 & 0 & 0.082 & 0 \\ 0 & 0 & 0 & 0 & 0 & 0 & 0 & 0.082 \end{bmatrix}$$

Similarly, the nondimensional material matrix $[\overline{D}]$ for cobalt poled relative to the toroidal φ axis with shell thickness h=0.1a_o is computed as,

$$[\overline{D}_\varphi] = \begin{bmatrix} 0.2892 & 0.0628 & 0 & 0 & 0 & 0 & 0 & 0 \\ 0.0628 & 0.4266 & 0 & 0 & 0 & 0 & 0 & 0 \\ 0 & 0 & 0.094 & 0 & 0 & 0 & 0 & 0 \\ 0 & 0 & 0 & 24.10(10^{-5}) & 5.23(10^{-5}) & 0 & 0 & 0 \\ 0 & 0 & 0 & 5.23(10^{-5}) & 35.55(10^{-5}) & 0 & 0 & 0 \\ 0 & 0 & 0 & 0 & 0 & 7.83(10^{-5}) & 0 & 0 \\ 0 & 0 & 0 & 0 & 0 & 0 & 0.082 & 0 \\ 0 & 0 & 0 & 0 & 0 & 0 & 0 & 0.082 \end{bmatrix}$$

The mass matrix is the same as that used in Section 3.

6.3. Comparison of Results

Frequency of vibration for thin shell analysis versus elasticity analysis is compared in Table 6.2. Shell thickness of h=0.1a_o is a thickness where good agreement can be expected when an elasticity analysis and a thin shell analysis are compared. The intent is to demonstrate the computational method and limited results are given in the table. The agreement between the two shell theories is satisfactory and indicates that the development of the various material matrices is correct.

Table 6.2. Comparison of frequencies □ $= \omega a_o\sqrt{\rho/C_{44}}$ for toroidal shells with toroidal radius R=5a_o, shell thickness h=019a_o based upon an elasticity analysis and a thin shell analysis using transversely isotropic material constants for cobalt and zinc

	cobalt, R=5a_o, h=0.1a_o						zinc, R=5a_o, h=0.1a_o	
	poled relative to r		poled relative to φ		poled relative to r		poled relative to φ	
n	elasticity	thin shell	elasticity	thin shell	elasticity	thin shell	elasticity	thin shell
0	0.1235	0.1207	0.1222	0.1184	0.1211	0.1173	0.1044	0.1016
	0.2732	0.2705	0.3075	0.3058	0.2840	0.2805	0.1877	0.1854
	0.3551	0.3538	0.4127	0.4105	0.3676	0.3669	0.2382	0.2324
1	0.1961	0.1939	0.1950	0.1891	0.2221	0.2190	0.1969	0.2075
	0.2822	0.2778	0.3036	0.2997	0.2854	0.2837	0.2217	0.2178
	0.3617	0.3609	0.4073	0.4024	0.4015	0.4027	0.2655	0.2659
2	0.0726	0.0727	0.0742	0.0736	0.0762	0.0762	0.0573	0.0584
	0.0763	0.0746	0.0761	0.0753	0.0768	0.0768	0.0595	0.0593
	0.2464	0.2443	0.2523	0.2472	0.2624	0.2615	0.2170	0.2174
3	0.1756	0.1741	0.1740	0.1714	0.1758	0.1767	0.1452	0.1501
	0.1833	0.1795	0.1793	0.1756	0.1786	0.1801	0.1513	0.1549
	0.2897	0.2888	0.3075	0.3012	0.3203	0.3230	0.2325	0.2393
4	0.2614	0.2588	0.2582	0.2545	0.2579	0.2620	0.2218	0.2278
	0.2655	0.2611	0.2609	0.2567	0.2592	0.2639	0.2237	0.2303
5	0.3425	0.3410	0.3425	0.3377	0.3438	0.3519	0.2750	0.2876
	0.3448	0.3417	0.3440	0.3386	0.3442	0.3526	0.2755	0.2881
6	0.4346	0.4332	0.4348	0.4276	0.4413	0.4531	0.3379	0.3592
	0.4359	0.4333	0.4359	0.4279	0.4415	0.4534	0.3382	0.3593

The analysis in axisymmetric cylindrical coordinates discussed in Section 5, while being very powerful for the study of toroidal shells with arbitrary shell cross section, becomes considerably more challenging for transversely isotropic materials. The surface of the shell is not defined by a coordinate and the material matrices that have been previously defined are not applicable. A method that can be used has been demonstrated by Liu and Buchanan (2006). The material matrix is transformed for every finite element to correspond to a coordinate direction that is normal to the shell surface. The transformation for a fourth order tensor, such as the material tensor, is as follows

$$C_{ijkl} = a_{ri}a_{sj}a_{tk}a_{ul}C'_{rstu} \tag{6.18}$$

where C_{ijkl} and C'_{rstu} are the transformed and original material tensors, respectively, and a_{pq} are transformation vectors. An in-depth analysis for a circular toroidal shell has been given by Liu and Buchanan (2006) and the interested reader should begin their study there.

CONCLUSION

The vibration of axisymmetric toroidal shells has been discussed and three different formulations for the mathematical description of the shell have been presented. The three formulations have in common that the finite element method was used to solve the governing differential equations. A finite element model valid for a thin shell with shear deformation and rotary inertia was developed in Section 2 that could be used for any shell of revolution. The only requirement is that the shell can be described in a coordinate system where Lamé parameters can be established. A complete non-axisymmetric shell of revolution or a sector of a shell could be modeled using a two-dimensional finite element. The equations of Section 2 were specialized to an axisymmetric toroidal shell of revolution in Section 3. Sufficient results were tabulated for frequency of free vibration for shells with free boundary conditions to illustrate the application of the methods of Section 3. A solution for vibration of thin toroidal shells including shear deformation and rotary inertia has previously not been recorded in the literature so it was necessary to verify the accuracy of the analysis. Fortunately, the method of analysis presented in Section 4 had been previously compared with existing literature and served to support and verify the thin shell analysis of Section 3. An analysis based upon the equations of elasticity cast in toroidal coordinates was developed and discussed in Section 4. The thin shell results of Section 3 were verified and additional analysis was reported for circular toroidal shells with a sector of the shell removed. These circular sectors were modeled assuming the shell was fixed at the boundaries where the sector was removed. The ability of the finite element method of analysis to model toroidal shells with an arbitrary cross-section was demonstrated in Section 5. The cross-section of the shell was assumed to lie in the (r,z) plane of an axisymmetric cylindrical coordinate system and finite element methods suitable for cylindrical coordinates were discussed. Results were presented for free vibration of toroidal shells with elliptic cross-section and free boundary conditions. In addition, a cross-section that appeared as an oblong section with spherical caps on the top and bottom was analyzed to demonstrate an application in cylindrical coordinates. The chapter ends with Section 6 and a discussion of toroidal shells with transversely isotropic material properties. It was shown that the thin shell analysis of Section 3 and the elasticity analysis of Section 4 could be cast in a format where material constants could represent transversely isotropic bodies. Results were given for a material that was transversely isotropic with respect to a direction normal to the surface of the shell. For contrast, the toroidal shell was assumed to be transversely isotropic relation to the φ coordinate that describes the circular direction around the shell. Results using the finite element of Sections 3 and 4 were compared and found to be in good agreement.

The FORTRAN codes used in the development of this chapter to solve the governing differential equations were user-written and all computations were performed on a personal computer with standard FORTRAN compiler and IMSL capability. Limited tabular results were presented. The chapter was intended to be informative and illustrative of methods for an analysis of toroidal shells and was not intended to be a research type report. Some new formulations and results were given that, to the author's knowledge, had not previously appeared in the literature. Again, the author would like to emphasize that the intent of the chapter is to give some detailed information for understanding the use of the finite element method in conjunction with several toroidal shell formulations. The reader is not expected to become an expert in code development or even detailed finite element concepts. But, an understanding of the concepts presented here will enable the reader to become more skilled with using a commercially available finite element code.

ACKNOWLEDGMENTS

The author is indebted to Dr. Jane Liu, Associate Professor of Civil and Environmental Engineering, Tennessee Technological University, for assistance with preparation of some of the figures. In addition, I appreciate the invitation extended by Dr. Bohua Sun to participate in the project of developing a volume dedicated to toroidal shells.

REFERENCES

Al-Khatib O. J., Buchanan G. R., Free vibration of a paraboloical shell of revolution including shear deformation and rotary inertia. *Thin-Wall. Struct.*, 2010, Vol. 48, 223-232.

Bisplinghoff R. L., Mar J. W., Pian T. H. H., *Statics of deformable bodies*. Addison-Wesley, Reading, MA, 1965, (reprinted by Dover).

Buchanan G. R., Liu Y. J., An analysis of the free vibration of thick-walled isotropic toroidal shells. *Int. J. Mech. Sci.,* 2005, Vol. 47, 277-292.

Buchanan G. R., *Theory and Problems of Finite Element Analysis*. Schaum's Outline Series, McGraw-Hill, New York, 1995.

Chou P. C., Pagano N. J., *Elasticity*. Van Nostrand, Princeton, NJ, 1967, (reprinted by Dover).

Hughes W. F., Gaylord E. W., *Basic Equations of Engineering Science*. Schaum Publishing Co., New York, NY,1964.

Hyer M. W., *Stress analysis of fiber-reinforced composite materials*. WCB/McGraw-Hill, Boston, MA, 1997.

Jiang W., Redekop D., Analysis of transversely isotropic hollow toroids using the semi-analytical DQM. *Struct. Eng. Mech.*, 2002, Vol. 13, 103-116.

Jones R. M., *Mechanics of composite materials*. Second edition, Taylor and Francis, Philadelphia, PA, 1999.

Liu Y. J., Buchanan G. R., Free vibration of a transversely isotropic thick-walled toroidal shell. *Int. J. Struct. Stab. Dyn.*, 2006, Vol.6, 359-376.

Liu Y. J., Madhavapeddy S., Buchanan G. R., Algebriac geometry approach in the modeling of a free vibration of laminated toroidal shells with elliptical cross-section. 43rd technical meeting of the Society of Engineering Science, August 2006, Penn State University.

Madhavapeddy S., *Free vibration analysis of complete isotropic toroidal shells with elliptical cross-section.* MS Thesis, Tennessee Technological University, Cookeville TN, USA, 2006.

Meirovitch L., *Analytical Methods in Vibrations*. Collier-MacMillan Lt., London, 1967.

Ming R. S., Pan J., Norton M. P., Free vibrations of elastic circular toroidal shells. *Appl. Acoust.*, 2002, Vol. 18, 513-528.

Nye J. F., *Physical properties of crystals.* Oxford University Press, 1969.

Payton R. G., *Elastic wave propagation in transversely isotropic media*, Martinus Nijhoff, The Hague, 1983.

Reddy J. N., *An Introduction to the Finite Element Method.* Second edition, McGraw-Hill, New York, NY, 1993.

Soedel W., *Vibrations of Shells and Plates*. Marcel Dekker, New York, NY, 1993.

Spiegel M. R., *Theory and Problems of Vector Analysis*. Schaum's Outline Series, McGraw-Hill, New York, NY, 1959.

Tessler A., An efficient, conforming axisymmetric shell element including transverse shear and rotary inertia. *Comput. Struct.,* 1982, Vol.15, 567-574.

Xu B., Redekop D., Natural frequencies of an orthotropic thin toroidal shell of elliptical cross-section. *J. Sound Vib.,* 2006, Vol. 293, 440-448.

Yamada G., Kogayashi Y., Ohta Y., Yokota S., Free vibration of a toroidal shell with elliptic cross-section. *J. Sound Vib.,* 1989, Vol. 135, 411-425.

Zhang F., Redekop D., Surface loading of a thin-walled toroidal shell. *Comput. Struct.,* 1992, Vol. 43, 1019-1104.

Zhou D., Au F. K. T., Lo S. H., Cheung Y. K., Three-dimensional vibration analysis of a torus with circular cross section. *J. Acoust. Soc. Amer.*, 2002, Vol. 112, 2831-2839.

In: Toroidal Shells
Editor: Bohua Sun

ISBN: 978-1-61942-247-6

Chapter 4

A FINITE ELEMENT FORMULATION FOR PIPING STRUCTURES BASED ON THIN SHELL DISPLACEMENTS THEORY[1]

E. M. M. Fonseca[1,*], F. J. M. Q. Melo[2] and L. R. Madureira[3]
[1]Department of Applied Mechanics, Ap.1134,
Polytechnic Institute of Bragança, Bragança, Portugal
[2]Department of Mechanical Engineering, University of Aveiro, Aveiro, Portugal
[3]Department of Mechanical Engineering,
Faculty of Engineering, Universidade do Porto, Portugal

NOMENCLATURE

The following symbols are used in this paper:

a_n; $\overline{a_n}$ = constants to be determined of the ovalization displacements;
$[B]$ = deformation matrix;
b_n; $\overline{b_n}$ = constants to be determined of the warping displacements;
$[D]$ = elasticity matrix;
E = elastic modulus;
$\{F_n\}$ = applied nodal mechanical forces;
$\{F_{th}\}$ = applied nodal thermal forces;
h = pipe thickness;
$[K]$ = stiffness matrix;

[1] This paper is based on the first author's postdoctoral research in piping structures. The research team has worked together in this field. Many publications have been published in different conferences and scientific journals.

* e-mail: efonseca@ipb.pt.

L= pipe length;
$M_{\theta\theta}$= circumferential stress;
N = shape function;
N_{ss} = longitudinal stress;
$N_{s\theta}$= shear stress;
R = mean curvature radius;
r = mean section pipe radius;
s = pipe arch length or longitudinal direction;
$[T]$= transpose matrix for global system;
U = tangential displacement;
u, v, w = shell surface displacements;
$u(s,\theta)$= longitudinal displacement from warping;
$v(s,\theta)$ = circumferential displacement from ovalization;
W = transverse displacement;
$\overline{W}$ = transverse displacement;
$w(s,\theta)$= radial displacement from ovalization;
α = central pipe angle;
β= rotation;
$\gamma_{s\theta}$= shear strain;
$\{\delta\}$= nodal displacement vector;
ε_{ss}= longitudinal strain;
$\varepsilon_{\theta\theta}$ = hoop strain;
ε^{mec} = mechanical strain;
ε^{th} = thermal strain;
θ = circumferential direction;
ν = Poisson's ratio;
φ = rotation;
$\overline{\varphi}$ = rotation;
$\chi_{\theta\theta}$ = circumferential curvature from ovalization;
$\overline{\alpha}$ = thermal expansion coefficient;
ΔT = temperature variation.

1. INTRODUCTION

The application of curved pipe elements is critically necessary in any piping structure, no matter which the project applications are. Depending on the safety standards ruling the project, it is known that curved pipe elements are much more vulnerable to excessive stressing than straight pipe elements. A larger risk of accessory breakdown during continuous operation after a situation where limiting forces are exceeded is likely to occur. The study of the flexibility and stress state of curved pipes subjected to generalized forces has been an area of interest of many engineers and physicists, given the high interest of the theme in many structural applications. Practically, in the beginning of 20th Century, important innovative contributions for the research was registered by Theodore von Karman (1911), who proposed the first really effective solution-based on a Fourier expansion displacement field and a variational method.

Vigness (1943) generalized the solution presented by Theodore von Karman (1911) and proposed an experimental procedure to verify the derived bending equations, Cheng and Thailer (1970) reduced the problem of in-plane bending to two coupled ordinary differential equations that could be integrated without simplifying assumptions.

G. Thomson (1980) worked with an analytical formulation using doubly-defined trigonometric series functions to develop the displacement field and performed many experimental studies, Öry and Wilczek (1983) presented an economical method based on transfer matrix techniques to define the stress and deformation calculation.

The technique of expanding the displacement and load field trigonometric functions has been largely followed until the emergence and continuous development of finite elements. This is a very powerful and flexible tool in numerical modelling of the many and various structural problems, with evident advantages in a more straightforward simulation of boundary conditions, non-linear behaviour and structure singularities. Many of these characteristics are hardly possible with Fourier expansion techniques, but the use of trigonometric functions in the approach to the solution of problems in structural mechanics has known encouraging contributions, when such trigonometric functions are combined with the current algebraic shape functions used in the development of finite elements [Ohtsubo and Watanabe (1978), K.J. Bathe and C.A. Almeida (1982)]. Numerical solutions for stress analysis of these elements were widely proposed by most researchers. Oñate (1992) presented a detailed solution based on double Fourier expansions for the displacement field. The study of the effect of radial loads on cylindrical shells has an important role in the engineering design and installation conditions of pressure vessels, having cylindrical shape. Evaluation methods of the structural effect of such loads have actually efficient numerical and experimental investigation tools. The first technique deals essentially with finite element techniques, actually based on continuous research work of enterprises as ABAQUS, NASTRAN, CAEPIPE, COSMOS/M, ANSYS and ALGOR. The experimental techniques have distinguished role, as it is possible to achieve realistic results, provided that the load conditions of the project should be reproduced as close as possible to the expected force system in the design to develop. An extended experimental analysis program using strain gauge techniques was performed by [Hose and Kitching (1993), Hughes and Kitching (1980)] where they experimentally studied the elastic behaviour of a cylindrical shell subjected to a radial load applied through two circumferential plate bracket attachments positioned diametrically opposite each other at the mid length of the cylinder. Rhodes (1997) describes an extensive list of contributions from researchers in the University of Strathclyde, where the stress state in cylindrical shells using experimental techniques has been investigated for several years. In this line of investigation and development, the researchers of this proposal have presented innovative solutions based on the formulation of finite ring elements [Melo and Castro (1992), (1997); Fonseca *et al*, (2002), (2005), (2006); Madureira *et al* (2000),

(2004)], exclusively designed for the deformation and stress analysis of curved pipes with generalized boundary conditions (rigid or thin flanges and tangent pipe terminations) and external forces. The pipe elements used a strain tensor combining a mixed mode deformation, where along the circumferential direction, the deformation model was assumed as the one of a Kirchhoff shell, while along the longitudinal direction (the pipe curvature radius), a deformation model of Reissner-Mindlin was considered. The non-linear deformation of pipes has also another important field of research, where the pipe material stays deformed in the elastic field, but the pipe geometry can present non-linear geometric deformations, either from large displacements resulting of bending moment intensity or a geometric instability from buckling, potentiated by local imperfections resulting of a lesser quality fabrication process. The pipe structure can present severe distortions under seismic disturbances when the structure experiences changes of boundary conditions or local constraints as result of loss of stability due to local buckling at cylindrical vessels attachments. In this work a computational model for stress analysis of tubular structures is presented. The work is based on shell displacement field using the elastic theory and a finite curved pipe element with two nodes is presented. The displacement field is based in trigonometric functions for the curved arch displacement and the development of Fourier series to model warping and ovalization of the tubular section. This element will be compared with results obtained with other element based in high order polynomial functions for the curved arch displacement, such as the equations developed by [Fonseca *et al* (2005), (2006)]. This element presents a good accuracy for the displacement and stress field resulting from in-plane and out-of-plane bending under generalized loads. With this element the behaviour of curved pipes end effects is also analyzed. The stress state calculation at any transverse section is easy, given the simplicity of the involved algorithms. The main advantage is associated to the generation of simple meshes with low number of elements, remarkable economy in degrees of freedom and in computational time.

2. Thin Shell Displacement Theory

In this section all the formulation and equations will be presented for piping structures based on thin shell displacements theory. The deformation field of a pipe element refers to membrane strain and curvature variation. The geometric parameters considered for this element definition are: the arc length (s), the mean radius of curvature (R), the thickness (h), the mean section radius of the pipe (r) and the central angle (α). Figure 2.1 shows the essential geometric parameters defining the two node finite curved pipe element (node i and j).

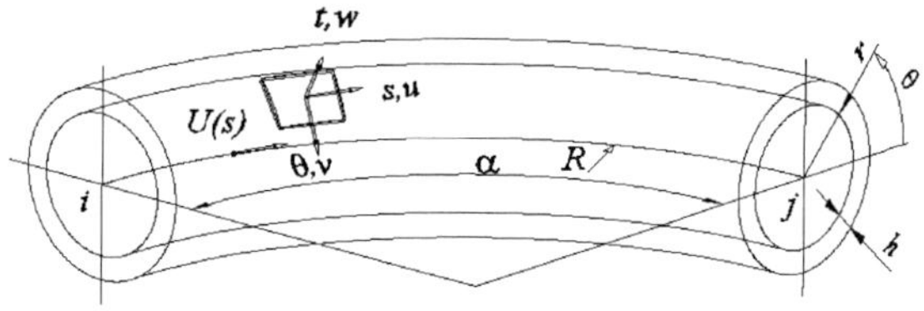

Figure 2.1. Geometric parameters for the finite curved pipe element.

2.1. Essential Assumptions

The basic equations governing the behaviour of thin shells were derived by Love (1944). The following assumptions referred in [Fonseca *et al* (2002), (2006), (2010)], were considered in the present analysis:

a) The curvature radius R is assumed much larger than the section radius r; this means that the pipe bore term $\left(R + r\cos\theta\right)$ may be approximated to R (figure 2.2);
b) A semi-membrane deformation model is adopted and neglects the bending stiffness along the longitudinal direction of the toroidal shell but considers the circumferential bending resulting from ovalization;
c) The shell is considered thin and inextensible along the circumferential direction θ, which means that the hoop strain $\varepsilon_{\theta\theta} = 0$, where:
d) The arch is thin, that is, $r \geq h$. Typically the thickness should be less than a tenth of the mean radius of the pipe.

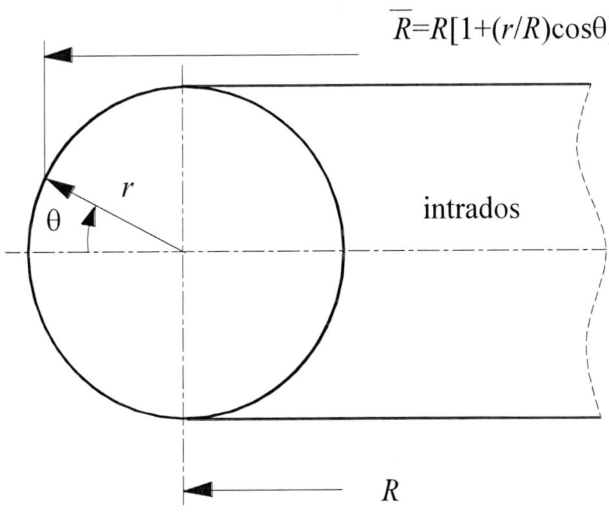

Figure 2.2. The radius of curvature for the curved pipe.

$$\varepsilon_{\theta\theta} = w + \frac{\partial v}{\partial \theta} \tag{2.1}$$

2.2. Thermal and Mechanical Formulation

The mechanical deformation model considers that the pipe undergoes a semi-membrane strain field, referred in [Fonseca *et al* (2006), (2010)], and also used by [Melo and Castro (1992), (1997); Flügge (1973); Kitching (1970)], and it is represented by the equation 2.2.

$$\underset{\sim mec}{\varepsilon} = \begin{Bmatrix} \varepsilon_{ss} \\ \gamma_{s\theta} \\ \chi_{\theta\theta} \end{Bmatrix} = \begin{bmatrix} \dfrac{\partial}{\partial s} & -\dfrac{\sin\theta}{R} & \dfrac{\cos\theta}{R} \\ \dfrac{1}{r}\dfrac{\partial}{\partial\theta} & \dfrac{\partial}{\partial s} & 0 \\ 0 & -\dfrac{1}{r^2}\dfrac{\partial}{\partial\theta} & \dfrac{1}{r^2}\dfrac{\partial^2}{\partial\theta^2} \end{bmatrix} \begin{Bmatrix} u \\ v \\ w \end{Bmatrix} \tag{2.2}$$

where: ε_{ss} is the longitudinal membrane strain, $\gamma_{s\theta}$ is the shear strain and $\chi_{\theta\theta}$ is the circumferential curvature form ovalization.

If a thermal deformation is considered, the dimension of the deformation vector increases one term due the thermal circumferential deformation by the expression:

$$\varepsilon_{\theta\theta} = \frac{1}{r}\left(w + \frac{\partial v}{\partial\theta}\right) = \frac{r\overline{\alpha}\Delta T}{r} \tag{2.3}$$

When a tubular system without restraint is subjected to temperature variation, there will be a length increase and the temperature produces dilation along the cross section of the pipe. The perimeter of the pipe will be variable. The thermal deformation vector is obtained from the following equation:

$$\begin{Bmatrix} \varepsilon_{ss} \\ \varepsilon_{\theta\theta} \end{Bmatrix}^{th} = \begin{Bmatrix} \overline{\alpha}\Delta T \\ \overline{\alpha}\Delta T \end{Bmatrix} \tag{2.4}$$

where $\overline{\alpha}$ is the thermal expansion coefficient and ΔT is the temperature variation.

Assuming small strains, the complete incremental relation between stress and strain for thermal and mechanical deformation is found to be:

$$\varepsilon = \varepsilon^{mec} + \varepsilon^{th} \tag{2.5}$$

where ε^{mec} is the mechanical strain increment and ε^{th} the thermal strain.

The application of the principle of virtual work finally gives the system of algebraic equations to be solved. The matrix force-displacement equation for this finite pipe element model is:

$$[K]\{\delta\} = \{F_n\} + \{F_{th}\} \tag{2.6}$$

where δ is a nodal unknown displacement vector, F_n is the applied nodal mechanical forces and F_{th} is a nodal force vector due to thermal effects.

The element stiffness matrix K is calculated from the matrix equation:

$$K_{global} = [T]\left(\int_S [B]^T [D][B] dS\right)[T]^T \tag{2.7}$$

where $dS = rdsd\theta$, T is the transpose matrix for global system, B results from the derivative of the shape functions for the finite pipe elbow element and the elasticity matrix D appears with a simple algebraic definition, dependent of the elastic modulus E, the pipe elbow thickness h and the Poisson's ratio ν.

In the pipe elbow element formulation, a Gaussian integration was carried out along variable s while an exact one was used along the circumferential direction θ.

In detail, the stiffness terms resulting from the beam shear deformation were calculated using one-point gauss integration, while all the remaining stiffness terms were calculated with two-point gauss integration (ovalization and warping terms). The total number of degrees of freedom for this element is $2(6+2N_\theta)$, where N_θ is the number of terms used in Fourier expansions (8 terms).

The following expression represents the nodal forces due to the thermal strain origin.

$$F_{th} = \int_V [B]^T [D]\{\varepsilon\}^{th} dV \tag{2.8}$$

where dV is the elementary pipe volume, D the elasticity matrix and ε^{th} the thermal strain.

Having solved the system of algebraic equations, the displacement field is calculated for all the nodes of the structure.

The stress field is then defined for each element in the following form:

$$\sigma = [D]\left(\{\varepsilon\}^{mec} - \{\varepsilon\}^{th}\right) + \{\sigma\}_0 \tag{2.9}$$

where σ_0 represents the initial stresses.

The elasticity matrix appears with a simple algebraic definition, where the off-diagonal terms vanish, according the following expression:

$$D = \begin{bmatrix} \frac{Eh}{1-\nu^2} & 0 & 0 \\ 0 & \frac{Eh}{2(1+\nu)} & 0 \\ 0 & 0 & \frac{Eh^3}{12(1-\nu^2)} \end{bmatrix} \tag{2.10}$$

2.3. Rigid Beam Displacement for a Two-Node Finite Pipe Elbow Element

The displacements *u*, *v* and *w* are calculated for the shell surface from the finite pipe elbow element (figure 2.1). These functions refer to the displacement field on the mean line arc (U , W and φ) as shown figure 2.3, for in-plane loading, designated as *IN*, where U is the tangential displacement, W the transverse displacement and φ the rotation.

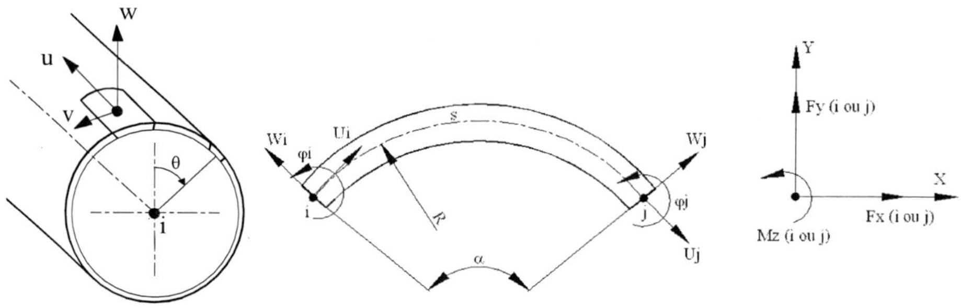

Figure 2.3. Degrees of freedom and applied forces and moments for in-plane element (*IN*).

For out-of-plane loading, designated as *OUT*, the displacement field is calculated also from the mean line of each arc considered like a rigid beam element: $\overline{W}$ is the transverse displacement, β and $\overline{\varphi}$ are the rotations in each direction represented in figure 2.4.

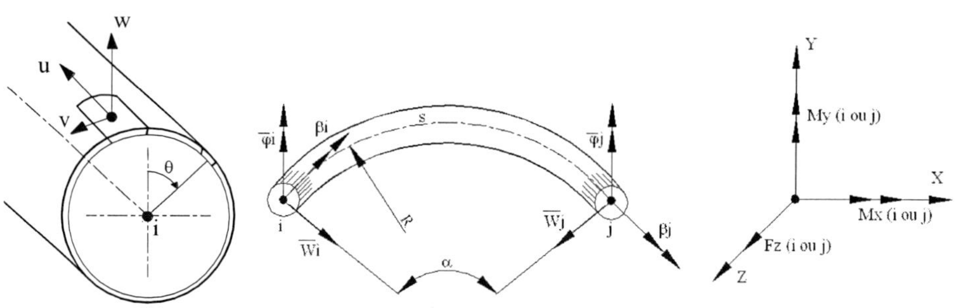

Figure 2.4. Degrees of freedom and applied forces or moments for out-of-plane element (*OUT*).

Those parameters are related *via* simple differential equations from beam bending theory, following simple hypotheses considered by [Melo and Castro (1992), (1997); G. Thomson (1980)]:

$$\varphi = \frac{dW}{ds} \tag{2.11}$$

$$W = -\frac{dU}{ds} R \tag{2.12}$$

2.4. Rigid Beam Displacement for in-Plane Loading

To find the shape functions for all degrees of freedom, when the finite pipe element has in-plane displacements (*IN*), a high order formulation is used and six parameters are necessary to define the displacement field. U can be approximated by the following fifth order polynomial (5P).

$$U_{(s)} = a_0 + a_1 s + a_2 s^2 + a_3 s^3 + a_4 s^4 + a_5 s^5 \tag{2.13}$$

The transverse displacement and the rotation can be calculated using equations 2.11 and 2.12:

$$W_{(s)} = -\frac{dU}{ds} R = -R\left(a_1 + 2a_2 s + 3a_3 s^2 + 4a_4 s^3 + 5a_5 s^4\right) \tag{2.14}$$

$$\varphi_{(s)} = \frac{dW}{ds} = -R\left(2a_2 + 6a_3 s + 12a_4 s^2 + 20a_5 s^3\right) \tag{2.15}$$

The coefficients in equations (2.13 at 2.15) are determined by imposing boundary conditions according to the curved reference.

With those specified conditions, the functions of the generic local displacements for an in-plane element are given by the following equations:

$$U_{(s)IN} = \left(U_i N_{ui} + U_j N_{uj}\right) + \left(W_i N_{wi} + W_j N_{wj}\right) + \left(\varphi_i N_{\varphi i} + \varphi_j N_{\varphi j}\right) \tag{2.16}$$

$$W_{(s)IN} = -R\left(\left(U_i N'_{ui} + U_j N'_{uj}\right) + \left(W_i N'_{wi} + W_j N'_{wj}\right) + \left(\varphi_i N'_{\varphi i} + \varphi_j N'_{\varphi j}\right)\right) \tag{2.17}$$

$$\varphi_{(s)IN} = -R\left(\left(U_i N''_{ui} + U_j N''_{uj}\right) + \left(W_i N''_{wi} + W_j N''_{wj}\right) + \left(\varphi_i N''_{\varphi i} + \varphi_j N''_{\varphi j}\right)\right) \tag{2.18}$$

The shape functions *N* are determined as follows, for each nodal element:

$$\begin{aligned} N_{ui} = {} & \cos\left(\frac{\alpha}{2}\right) + \frac{1}{R}\sin\left(\frac{\alpha}{2}\right) s + \left(-\frac{10}{L^3}\cos\left(\frac{\alpha}{2}\right) - \frac{6}{RL^2}\sin\left(\frac{\alpha}{2}\right)\right) s^3 + \\ & + \left(\frac{15}{L^4}\cos\left(\frac{\alpha}{2}\right) + \frac{8}{RL^3}\sin\left(\frac{\alpha}{2}\right)\right) s^4 + \left(-\frac{6}{L^5}\cos\left(\frac{\alpha}{2}\right) - \frac{3}{RL^4}\sin\left(\frac{\alpha}{2}\right)\right) s^5 \end{aligned} \tag{2.19}$$

$$\begin{aligned} N_{uj} = {} & \left(\frac{10}{L^3}\cos\left(\frac{\alpha}{2}\right) + \frac{4}{RL^2}\sin\left(\frac{\alpha}{2}\right)\right) s^3 + \left(-\frac{15}{L^4}\cos\left(\frac{\alpha}{2}\right) - \frac{7}{RL^3}\sin\left(\frac{\alpha}{2}\right)\right) s^4 + \\ & + \left(\frac{6}{L^5}\cos\left(\frac{\alpha}{2}\right) + \frac{3}{RL^4}\sin\left(\frac{\alpha}{2}\right)\right) s^5 \end{aligned} \tag{2.20}$$

$$N_{wi} = \sin\left(\frac{\alpha}{2}\right) - \frac{1}{R}\cos\left(\frac{\alpha}{2}\right)s + \left(-\frac{10}{L^3}\sin\left(\frac{\alpha}{2}\right) + \frac{6}{RL^2}\cos\left(\frac{\alpha}{2}\right)\right)s^3 + \\ + \left(\frac{15}{L^4}\sin\left(\frac{\alpha}{2}\right) - \frac{8}{RL^3}\cos\left(\frac{\alpha}{2}\right)\right)s^4 + \left(-\frac{6}{L^5}\sin\left(\frac{\alpha}{2}\right) + \frac{3}{RL^4}\cos\left(\frac{\alpha}{2}\right)\right)s^5 \tag{2.21}$$

$$N_{wj} = \left(-\frac{10}{L^3}\sin\left(\frac{\alpha}{2}\right) + \frac{4}{RL^2}\cos\left(\frac{\alpha}{2}\right)\right)s^3 + \left(\frac{15}{L^4}\sin\left(\frac{\alpha}{2}\right) - \frac{7}{RL^3}\cos\left(\frac{\alpha}{2}\right)\right)s^4 + \\ + \left(-\frac{6}{L^5}\sin\left(\frac{\alpha}{2}\right) + \frac{3}{RL^4}\cos\left(\frac{\alpha}{2}\right)\right)s^5 \tag{2.22}$$

$$N_{\varphi i} = -\frac{1}{2R}s^2 + \frac{3}{2LR}s^3 - \frac{3}{2L^2R}s^4 + \frac{1}{2L^3R}s^5 \tag{2.23}$$

$$N_{\varphi j} = -\frac{1}{2LR}s^3 + \frac{1}{L^2R}s^4 - \frac{1}{2L^3R}s^5 \tag{2.24}$$

An alternative solution was developed using a formulation based on trigonometric functions (TF). This solution uses four parameters to define the displacement field.

U can be approximated by the following function:

$$U_{(s)} = a_1\cos\left(\frac{s}{R}\right) + a_2\sin\left(\frac{s}{R}\right) + a_3\cos\left(2\frac{s}{R}\right) + a_4\sin\left(2\frac{s}{R}\right) \tag{2.25}$$

The transverse displacement refers to the inextensibility for arch mean curved line and can be calculated:

$$W_{(s)} = -R\frac{dU}{ds} = a_1\sin\left(\frac{s}{R}\right) - a_2\cos\left(\frac{s}{R}\right) + a_3 2\sin\left(2\frac{s}{R}\right) - a_4 2\cos\left(2\frac{s}{R}\right) \tag{2.26}$$

For the rotation field a linear polynomial is defined from an uncoupled equation:

$$\varphi_{(s)} = a_5 + a_6 S \tag{2.27}$$

2.5. Rigid Beam Displacement for Out-of-Plane Loading

The out-of-plane displacement field, in a local reference system, is determined with the following equations:

$$W_{(s)OUT} = \overline{W}_i N_1 - \overline{\varphi}_i N_2 + \overline{W}_j N_3 - \overline{\varphi}_j N_4 \tag{2.28}$$

$$\phi_{(s)OUT} = \frac{\partial W}{\partial s} = \overline{W}_i N_1' - \overline{\phi}_i N_2' + \overline{W}_j N_3' - \overline{\phi}_j N_4' \quad (2.29)$$

$$\beta_{(s)OUT} = \beta_i N_i + \beta_j N_j \quad (2.30)$$

The shape functions used in equation (2.28, 2.29, 2.30) are third order polynomials (N_1, N_2, N_3, N_4) and first order polynomials (N_i, N_j), respectively as following:

$$N_1 = 1 - \frac{3s^2}{L^2} + \frac{2s^3}{L^3} \quad (2.31)$$

$$N_2 = s - \frac{2s^2}{L} + \frac{s^3}{L^2} \quad (2.32)$$

$$N_3 = \frac{3s^2}{L^2} - \frac{2s^3}{L^3} \quad (2.33)$$

$$N_4 = -\frac{s^2}{L} + \frac{s^3}{L^2} \quad (2.34)$$

$$N_i = 1 - \frac{s}{L} \quad (2.35)$$

$$N_j = \frac{s}{L} \quad (2.36)$$

2.6. Surface Displacement Resulting from Ovalization and Warping

The surface displacement in the radial direction resulting only from the pipe ovalization, for in-plane and out-of-plane, according G. Thomson (1980), is expressed by the equation:

$$w(s,\theta) = \left(\sum_{n\geq 2} a_n \cos n\theta + \sum_{n\geq 2} \overline{a}_n \sin n\theta\right) N_i + \left(\sum_{n\geq 2} a_n \cos n\theta + \sum_{n\geq 2} \overline{a}_n \sin n\theta\right) N_j \quad (2.37)$$

The circumferential displacement, also resulting from the ovalization, is calculated from the following equation:

$$v(s,\theta) = \left(-\sum_{n\geq 2} \frac{a_n}{n} \sin n\theta + \sum_{n\geq 2} \frac{\overline{a}_n}{n} \cos n\theta\right) N_i + \left(-\sum_{n\geq 2} \frac{a_n}{n} \sin n\theta + \sum_{n\geq 2} \frac{\overline{a}_n}{n} \cos n\theta\right) N_j \quad (2.38)$$

Finally, the longitudinal displacement as a consequence of the transverse pipe section warping is defined from the following equation:

$$u(s,\theta)=\left(\sum_{n\geq 2} b_n \cos n\theta+\sum_{n\geq 2} \bar{b}_n \sin n\theta\right)N_i+\left(\sum_{n\geq 2} b_n \cos n\theta+\sum_{n\geq 2} \bar{b}_n \sin n\theta\right)N_j \tag{2.39}$$

The terms a_n and $\bar{a}_n$ are constants to be determined and included in the Fourier expansions for the ovalization displacements of in-plane and out-of-plane bending, respectively. The parameters b_n and $\bar{b}_n$ are also functions of developed series resulting from warping displacements in and out-of-plane. These terms are vectors and involving edge terms for the pipe ovalization and warping.

2.7. Shell Displacement Field for in and Out-of Plane Loading

The developed finite-element has two nodal sections and a total of 22 degrees of freedom (3 translations, 3 rotations, 8 terms for ovalization and 8 terms for warping), for in and out-of-plane loading. This displacement field is presented according the following vector, and used in equation 2.6:

$$\{\delta\}_{mec(IN+OUT)}=\left\{\underbrace{U\ W\ \bar{W}\ \beta\ \bar{\phi}\ \phi\ a_n\ \bar{a}_n\ b_n\ \bar{b}_n}_{i}\ \underbrace{U\ W\ \bar{W}\ \beta\ \bar{\phi}\ \phi\ a_n\ \bar{a}_n\ b_n\ \bar{b}_n}_{j}\right\}^T \tag{2.40}$$

The proposed formulation for the shell finite-element displacement field, resulting from the superposition of rigid beam displacement, a complete Fourier expansion for ovalization and warping terms, and the presence of temperature variation, for in and out-of-plane loading, and leads to:

$$u=U_{(s)_{IN}}-r\cos\theta\varphi_{(s)_{IN}}-r\sin\theta\varphi_{(s)_{OUT}}+u(s,\theta)+s\bar{\alpha}\Delta T \tag{2.41}$$

$$v=-W_{(s)_{IN}}\sin\theta+W_{(s)_{OUT}}\cos\theta+r\beta_{(s)_{OUT}}+v(s,\theta) \tag{2.42}$$

$$w=W_{(s)_{IN}}\cos\theta+W_{(s)_{OUT}}\sin\theta+w(s,\theta)+r\bar{\alpha}\Delta T \tag{2.43}$$

In these equations, $w(s,\theta)$ is the surface displacement in radial direction results from ovalization resulting from equation 2.39; $v(s,\theta)$ the circumferential displacement due to ovalization, equation 2.38; $u(s,\theta)$ is the longitudinal displacement due to warping tubular section effect, equation 2.37; U is the tangential beam displacement, W the transversal beam displacement and φ represent the beam rotation in z direction, all these resulting from in and out-of-plane loading as presented in previous chapters.

The displacement field of the shell surface in a condensed vector, due any mechanical or thermal loading conditions, is represented:

$$\begin{Bmatrix} u \\ v \\ w \end{Bmatrix} = \begin{Bmatrix} u \\ v \\ w \end{Bmatrix}_{mec(IN+OUT)} + \begin{Bmatrix} u \\ v \\ w \end{Bmatrix}_{th} = [N]\times\{\delta\}_{mec(IN+OUT)} + \begin{Bmatrix} s\bar{\alpha}\Delta T \\ 0 \\ r\bar{\alpha}\Delta T \end{Bmatrix} \tag{2.44}$$

where $\bar{\alpha}$ is the thermal expansion coefficient, considered constant in this formulation, ΔT is the temperature variation, N the shape functions used, and δ the displacements.

According to all formulation presented in this work, the finite shell element displacement for in-plane loading can be presented in the following algebraic matrix:

$$\begin{Bmatrix} u \\ v \\ w \end{Bmatrix}_{mec(IN)} = \begin{bmatrix} N_{ui} + RN''_{ui} r\cos\theta & N_{wi} + RN''_{wi} r\cos\theta & N_{\phi i} + RN''_{\phi i} r\cos\theta & 0 & \underset{\sim}{\cos i\theta N_i} \\ RN'_{ui}\sin\theta & RN'_{wi}\sin\theta & RN'_{\phi i}\sin\theta & -\underbrace{\frac{\sin n\theta}{n}}_{\sim} N_i & 0 \\ -RN'_{ui}\cos\theta & -RN'_{wi}\cos\theta & -RN'_{\phi i}\cos\theta & \underbrace{\cos n\theta}_{\sim} N_i & 0 \end{bmatrix}$$

$$\begin{bmatrix} N_{uj} + RN''_{uj} r\cos\theta & N_{wj} + RN''_{wj} r\cos\theta & N_{\phi j} + RN''_{\phi j} r\cos\theta & 0 & \underbrace{\cos n\theta}_{\sim} N_j \\ RN'_{uj}\sin\theta & RN'_{wj}\sin\theta & RN'_{\phi j}\sin\theta & -\underbrace{\frac{\sin n\theta}{n}}_{\sim} N_j & 0 \\ -RN'_{uj}\cos\theta & -RN'_{wj}\cos\theta & -RN'_{\phi j}\cos\theta & \underbrace{\cos n\theta}_{\sim} N_j & 0 \end{bmatrix} \times \begin{Bmatrix} U_i \\ W_i \\ \phi_i \\ \underset{\sim}{a_{ni}} \\ \underset{\sim}{b_{ni}} \\ U_j \\ W_j \\ \phi_j \\ \underset{\sim}{a_{nj}} \\ \underset{\sim}{b_{nj}} \end{Bmatrix} \tag{2.45}$$

Finally, the finite shell element displacement for out-of-plane loading mechanical condition is presented in the following algebraic system:

$$\begin{Bmatrix} u \\ v \\ w \end{Bmatrix}_{mec(OUT)} = \begin{bmatrix} -N'_1 r\sin\theta & 0 & N'_2 r\sin\theta & 0 & \underset{\sim}{\sin n\theta} N_i \\ N_1\cos\theta & rN_i & -N_2\cos\theta & \underbrace{\frac{\cos n\theta}{n}}_{\sim} N_i & 0 \\ N_1\sin\theta & 0 & -N_2\sin\theta & \underbrace{\sin n\theta}_{\sim} N_i & 0 \end{bmatrix}$$

$$\times \begin{bmatrix} -N'_3 r\sin\theta & 0 & N'_4 r\sin\theta & 0 & \sin n\theta N_j \\ N_3\cos\theta & rN_j & -N_4\cos\theta & \underbrace{\frac{\cos n\theta}{n}}_{\sim} N_j & 0 \\ N_3\sin\theta & 0 & -N_4\sin\theta & \underbrace{\sin n\theta}_{\sim} N_j & 0 \end{bmatrix} \times \begin{Bmatrix} \overline{W}_i \\ \beta_i \\ \overline{\phi}_i \\ \underset{\sim}{\overline{a}_{in}} \\ \underset{\sim}{\overline{b}_{in}} \\ \overline{W}_j \\ \beta_j \\ \overline{\phi}_j \\ \underset{\sim}{\overline{a}_{nj}} \\ \underset{\sim}{\overline{b}_{nj}} \end{Bmatrix} \tag{2.46}$$

3. Case Studies

To assess the accuracy of the pipe element in discussion, some examples are analyzed in this section. In some of them, the external loads are conventional in the project of piping systems (as for example, the pure bending moment) while in other, less conventional external loads were included in the examples. Apparently, they have less interest in the design of piping systems and in fact they are avoided, as is the example of edge concentrated loads. Their inclusion in this work is associated with a more elaborated solution needed for the stress analysis, hence their presentation here. All forces and moments considered must be referred to the global system of the element, except constriction forces due to ovalization or warping displacements.

3.1. Curved Pipe under in-Plane Pure Bending

The following example refers to a curved pipe with end restraints subjected to a uniform bending moment, as shown in figure 3.1. This example was taken from the work of Wilczek (1984) who investigated the stress analysis of curved pipes using the transfer matrix method and experimental tests. This example was also analyzed by Melo and Castro (1992), (1997) with a finite pipe element developed by them. Here only one half of the pipe bend was analyzed due to geometry and loading symmetry. The results refer to the transverse section equidistant from the pipe ends, either with rigid or thin flanges, as shown in figure 3.1.

Figures 3.2 and 3.3 present the longitudinal stresses of a curved pipe under pure bending moment, using rigid or thin flanges. The results were obtained with the presented formulation based on trigonometric functions (TF) for the beam terms or using a fifth order polynomial (5P), with different meshes, and were compared with the results produced by [Melo and Castro (1992), (1997), and Wilczek (1984)]. The longitudinal stresses are represented along the half span (0º–180º) of the transverse pipe section, for each type of applied flanges.

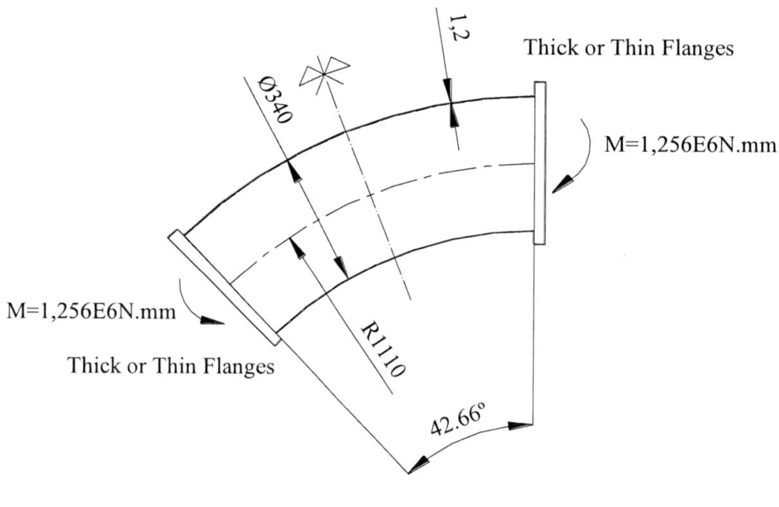

Figure 3.1. Geometry of a curved pipe loaded by a concentrated moment with end restraints.

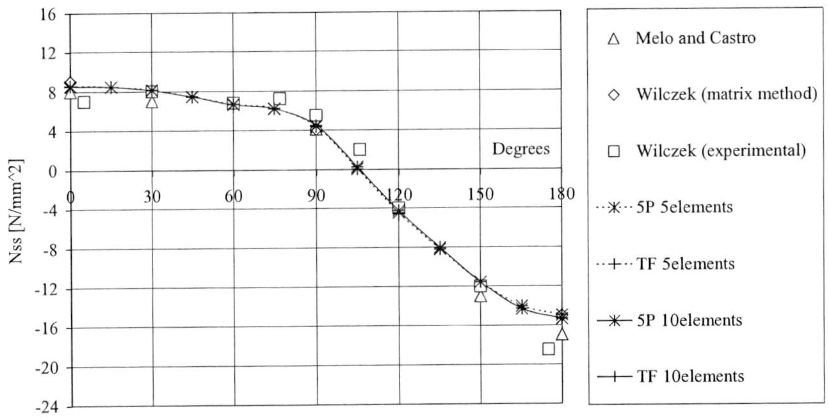

Figure 3.2. Longitudinal stresses for a curved pipe with rigid flanges.

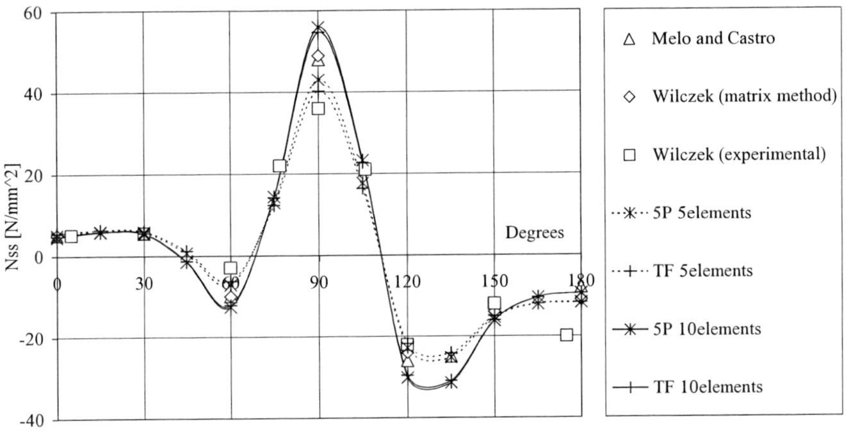

Figure 3.3. Longitudinal stresses for a curved pipe with thin flanges.

The presented results with the finite pipe element formulation, agree with other presented references. Also the numerical results with lower meshes present a good approximation to experimental results from other authors. The results obtained with thin flanges present higher value stresses than when rigid flanges are applied. The figure 3.3 shows a typical graphic with this type of behaviour. At the middle of pipe section (90°) the stresses are in tensile and have a peak value.

3.2. Curved Pipe Loaded with a Pair of Concentrated Forces

Figure 3.4 presents the same component represented in figure 3.1, but now loaded with a bending moment resulting from a pair of concentrated forces, acting on a thin flange. The flange of the built-in edge is considered rigid. Thin flange does not ovalize and the warp forces are considered like as referred by Melo and Castro (1992), (1997).

For a pair of concentrated longitudinal forces, of opposed senses, acting like an in-plane bending moment, the force vector must be obtained with the following expression:

$$\{F_n\}^T = \left\{ \underbrace{0,0,2F \times r}_{beam}, \underbrace{0,0,0,0,0,0,0,0}_{ovalization}, \underbrace{0,2F,0,2F,0,2F,0,2F}_{warping} \right\}^T \tag{3.1}$$

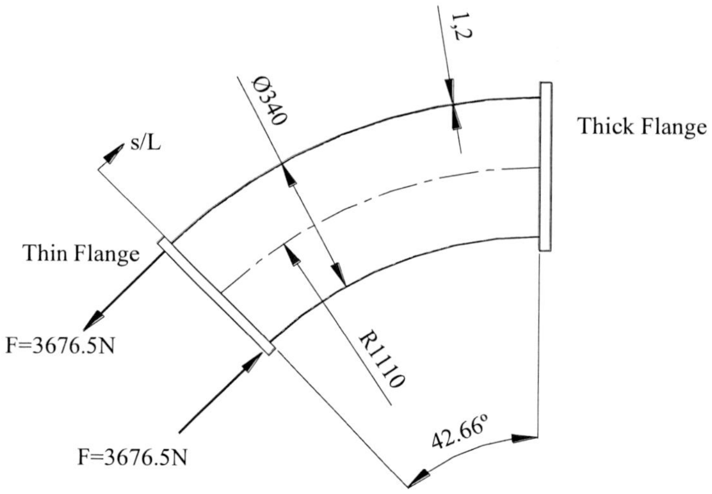

Figure 3.4. Geometry of a curved pipe loaded by a pair of concentrated forces.

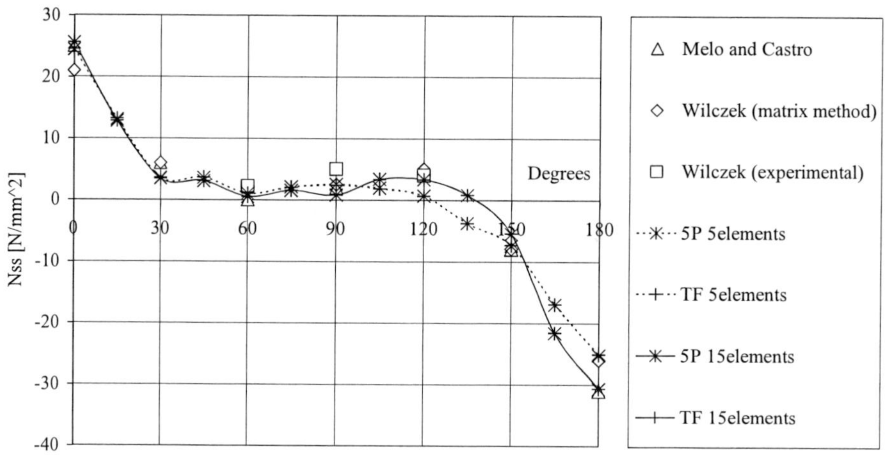

Figure 3.5. Longitudinal stress for a curved pipe loaded by a pair of forces, (s/L=0,1).

Figures 3.5, 3.6 and 3.7 represent the longitudinal membrane stresses for three different sections positions, along the longitudinal direction of the curved pipe. For the obtained results two different mesh refinements were used. The results were compared with corresponding data from other references.

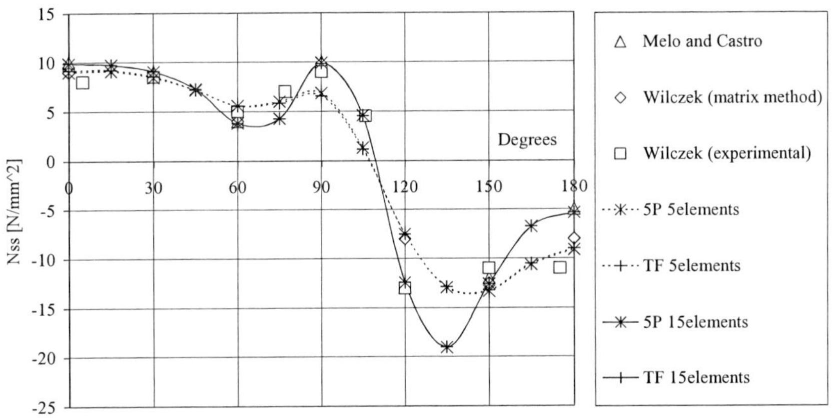

Figure 3.6. Longitudinal stress for a curved pipe loaded by a pair of forces, (s/L=0,5).

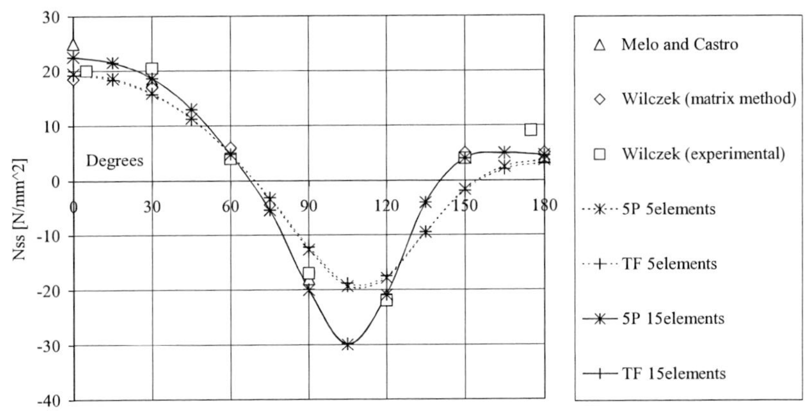

Figure 3.7. Longitudinal stress for a curved pipe loaded by a pair of forces, (s/L=0,9).

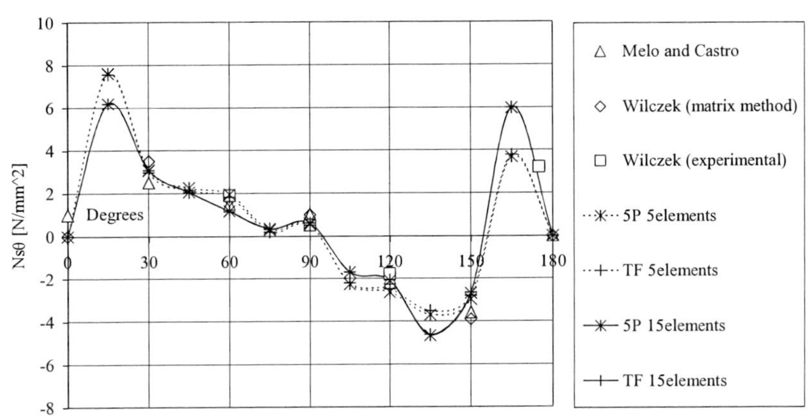

Figure 3.8. Shear membrane stress for a curved pipe loaded by a pair of forces, (s/L=0,1).

Good agreement between results was observed, although less accuracy was obtained for results in the section close to the built-in edge, when a coarse mesh was used.

Figures 3.8, 3.9 and 3.10 represent the shear membrane stresses for the same different positions on the curved pipe length.

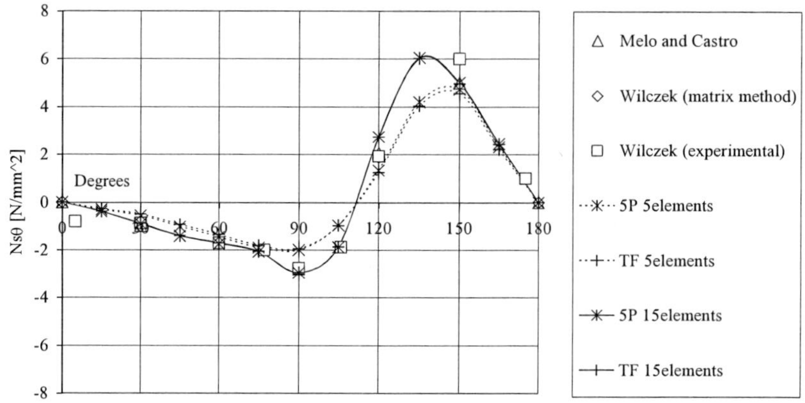

Figure 3.9. Shear membrane stress for a curved pipe loaded by a pair of forces, (s/L=0,5).

Also a good approximation even when a coarse mesh is used when compared with the results calculated by [Wilczek (1984), Melo and Castro (1992), (1997)]. The results of shear membrane stress near at the built-end are worst, when lower elements were used in the formulation.

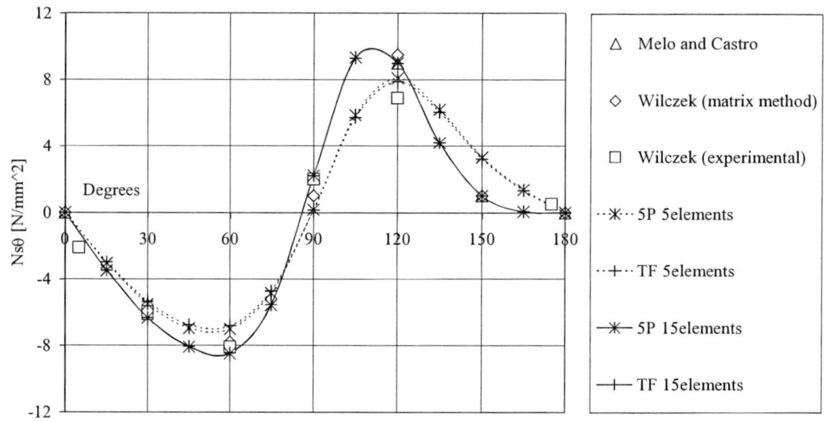

Figure 3.10. Shear membrane stress for a curved pipe loaded by a pair of forces, (s/L=0,9).

3.3. A Pipe Elbow System Subjected to Axial Forces

The next studied case is shown in figure 3.11, representing a tubular steel structure submitted to an axial force in their extremities. The elasticity modulus is 2.1GPa and the Poisson coefficient is equal to 0.3.

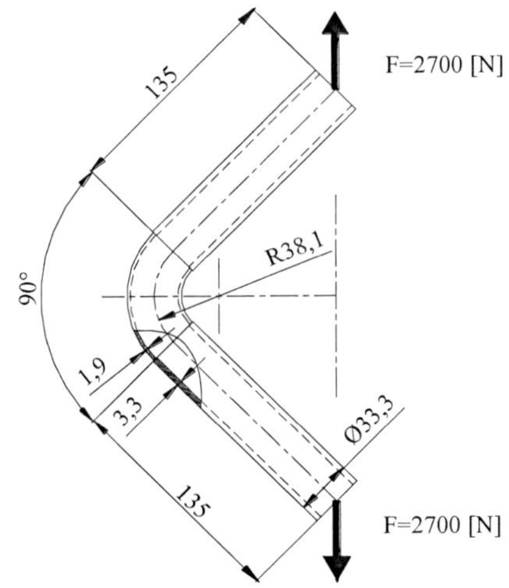

Figure 3.11. Geometric parameters for a pipe elbow structure.

An experimental setup was used for determining the longitudinal stresses at the middle of the structure over the radial cross section, using an experimental strain gauge method. The loading system was implemented by the universal machine as represented in the figure 3.12.

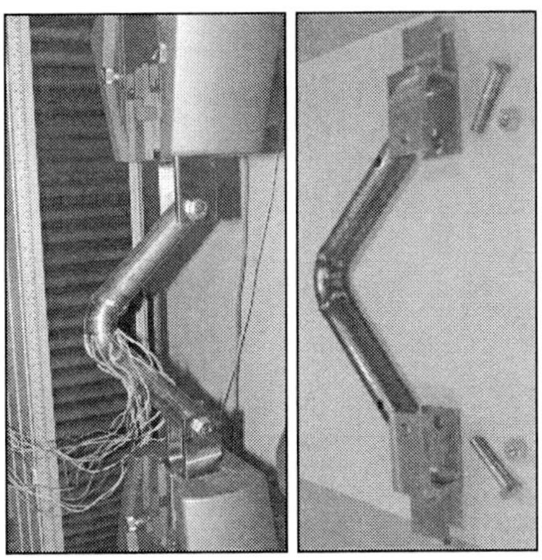

Figure 3.12. Experimental assembly for a pipe elbow structure.

The longitudinal stresses will be analysed using the experimental strain gauge method and compared with the numerical results obtained with ANSYS program and with the finite pipe elbow element. In the ANSYS program, two different meshes were used, with Shell63 element. Figure 3.13 shows the stresses field in the tubular structure.

For numerical results with the finite pipe elbow element two different meshes were used, as represented in figure 3.14. The stresses results were obtained for the transverse section at the middle of the structure.

The results of longitudinal stresses at the middle of the pipe elbow structure are represented in figure 3.15.

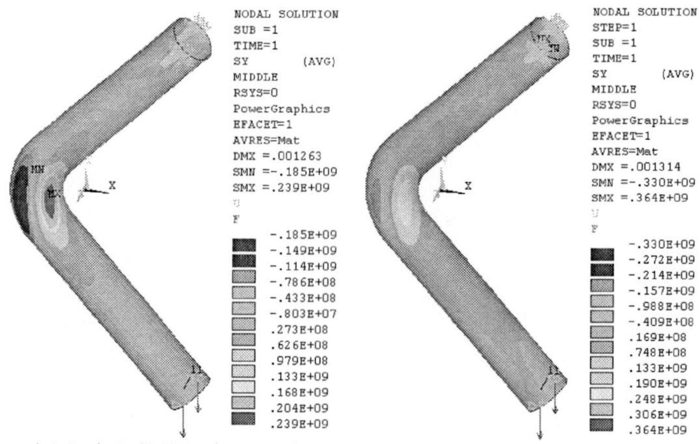

Mesh 1 (624 elements) Mesh 2 (2688 elements).

Figure 3.13. Stresses field using ANSYS program.

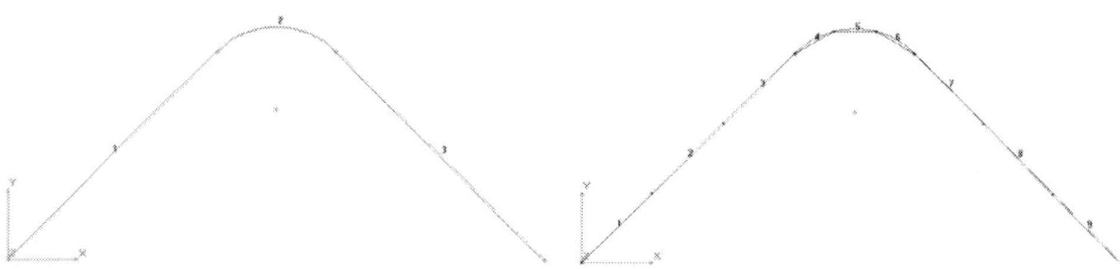

3 elements 9 elements.

Figure 3.14. One-dimensional meshes used in developed finite element.

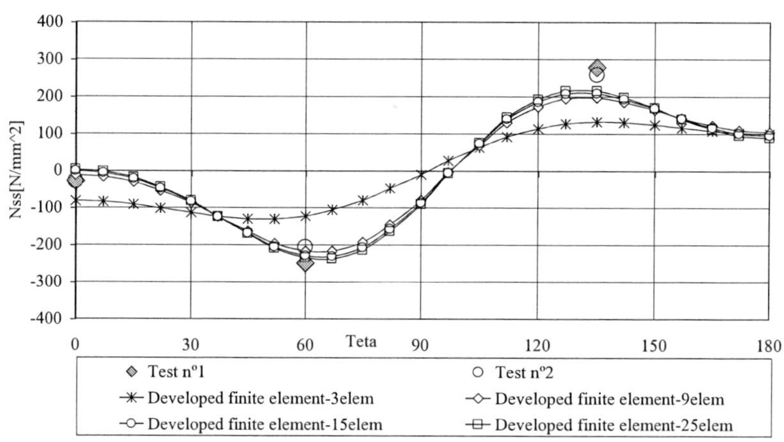

Figure 3.15. Longitudinal stresses with developed finite element and experimental results.

The numerical results using 9 elements present a good agreement with the experimental results. Using other higher mesh density, the results converge to the same final solution, as represented in figure 3.15.

Figure 3.16 presents the results obtained with the developed pipe elbow finite element, with the experimental results and with ANSYS program.

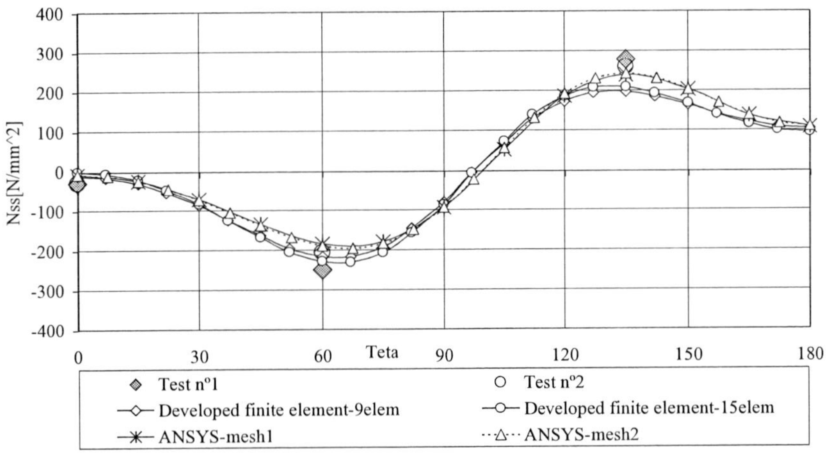

Figure 3.16. Longitudinal stresses with finite element, experimental results and ANSYS program.

This case study presents a methodology for stresses characterization in pipe elbow structures. One experimental in-plane loading pipe elbow system was presented to validate the numerical procedures presented. The finite pipe elbow element gives excellent results compared with measured stresses and numerical shell model. The experimental results were determined over the external tubular structure surface, justifying the position in figures 3.15 and 3.16.

3.4. A Pipe Elbow System Subjected to Elevated Temperatures

A tubular steel pipe elbow system has a radius of curvature equal 0.25m, a mean radius of 0.0135m and the thickness is 0.001m, figure 3.17. The piping system has end restraints and is subjected to different temperatures. The thermal expansion coefficient $\overline{\alpha}$ is constant and equal to $14x10^{-6}$ $°C^{-1}$, $\nu = 0.3$ and E is temperature dependent, according Eurocode3 (1995).

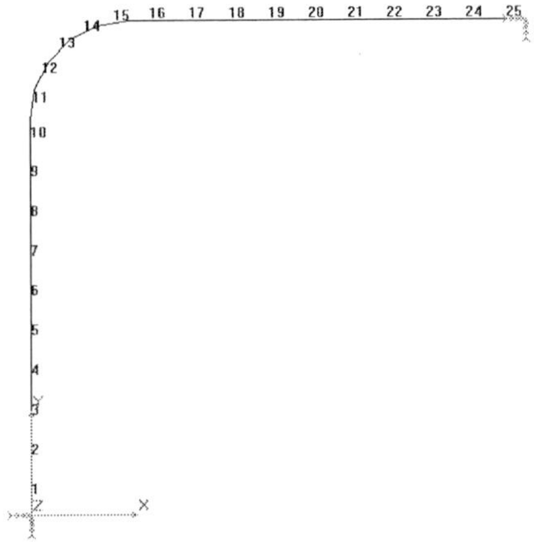

Figure 3.17. One-dimensional mesh used with finite pipe elbow element.

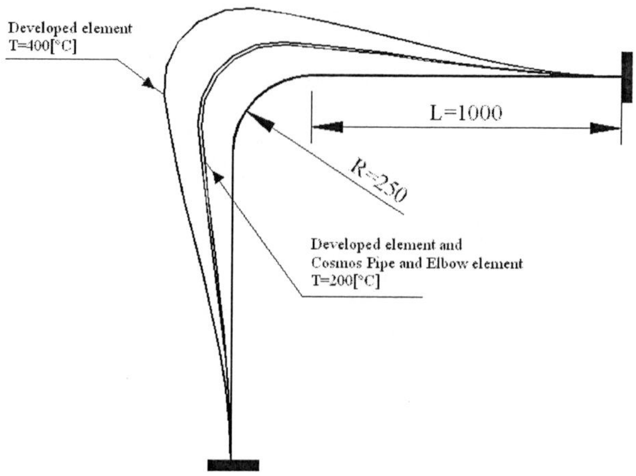

Figure 3.18. Displacement results for different imposed temperatures.

Figure 3.18 represents the deformed shape obtained and magnified 30 times. For a temperature rise of T=200°C numerical results have been compared with a commercial program using *Pipe* and *Elbow* elements. Similar results were obtained and compared. With temperature increase the pipe elbow system presents higher deformed shape at the mean elbow zone.

Figures 3.19 and 3.20 represent the transverse displacement obtained with different temperatures for a medium section in the tubular straight pipe, element number 5 in figure 3.17.

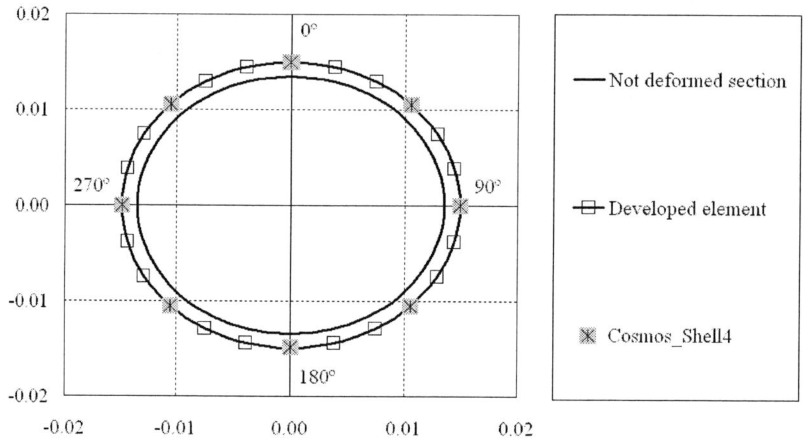

Figure 3.19. Pipe cross-section expansion using the finite-element and COSMOS (T=200°C).

As can be observed in figures 3.19 and 3.20, the pipe cross-section has the mean radius increased with the thermal expansion. The results obtained with the element developed herein are compared with the COSMOS program using a *Shell* element. The use of pipe elbow element formulation is an advantage; because it is possible with a simple element to calculate the displacement field in surface shell, without using expensive meshes.

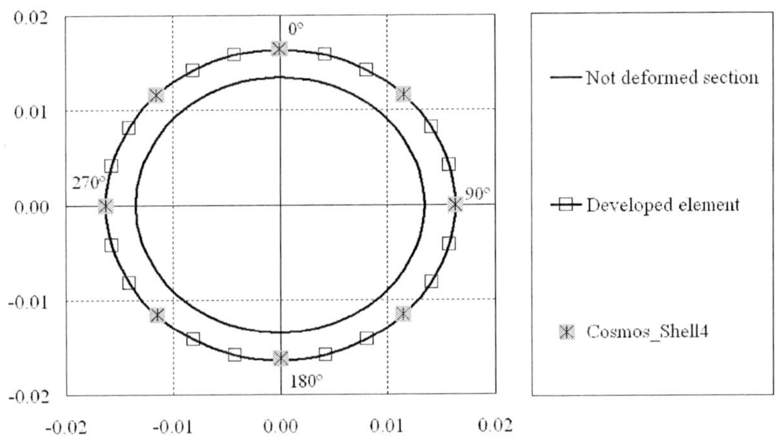

Figure 3.20. Pipe cross-section expansion using the finite-element and COSMOS (T=400°C).

3.5. A Spatial Pipe Elbow System Subjected to Elevated Temperatures and a Vertical Force

The next case is a structural steel pipe elbow system with all elements subjected to uniform temperature (T=200°C or T=600°C) and a vertical force of F=3000N at the mid-length of the structure, as shown in figure 3.21. The structure has end restraints. The system has a mean radius of 0.022m and 0.0025m of thickness. The thermal expansion coefficient considered is equal to 14x10^{-6} °C^{-1}, $\nu = 0.3$ and E is function of temperature, according Eurocode3 (1995). The finite pipe elbow element mesh used has 65 elements, as represented in figure 3.21.

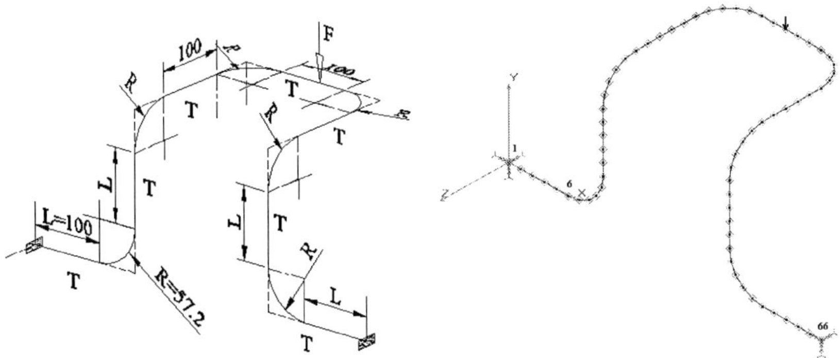

Figure 3.21. Pipe elbow system, with uniform temperature and vertical force.

The results from transversal displacements, figures 3.22 at 3.27, are compared with those obtained using the COSMOS programme with *Pipe* and *Elbow* elements. Also the temperature influence in the structure was compared simultaneously with or without a vertical force in the middle of system.

The longitudinal displacement U is higher when increasing the temperature, figures 3.22 and 3.23. As one can see, the vertical force has lower influence when compared with temperature.

Figure 3.22. Displacement U using the pipe elbow element and COSMOS program (T=200°C).

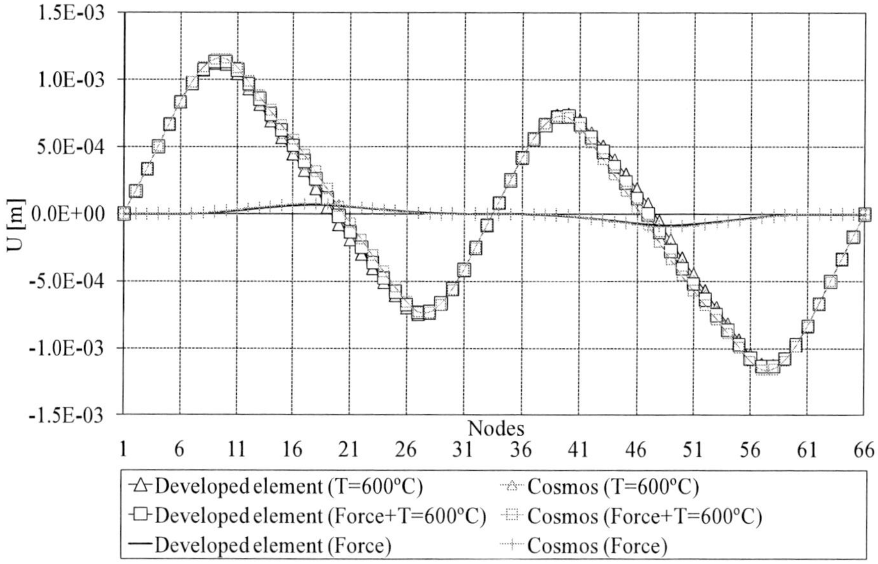

Figure 3.23. Displacement U using the pipe elbow element and COSMOS program (T=600°C).

The transversal displacement W is also higher when increase the temperature, figures 3.24 and 3.25. For this type of displacement the vertical force has influence in the results, mainly close to their application. The vertical force decreases the higher values of transversal displacement obtained with temperature.

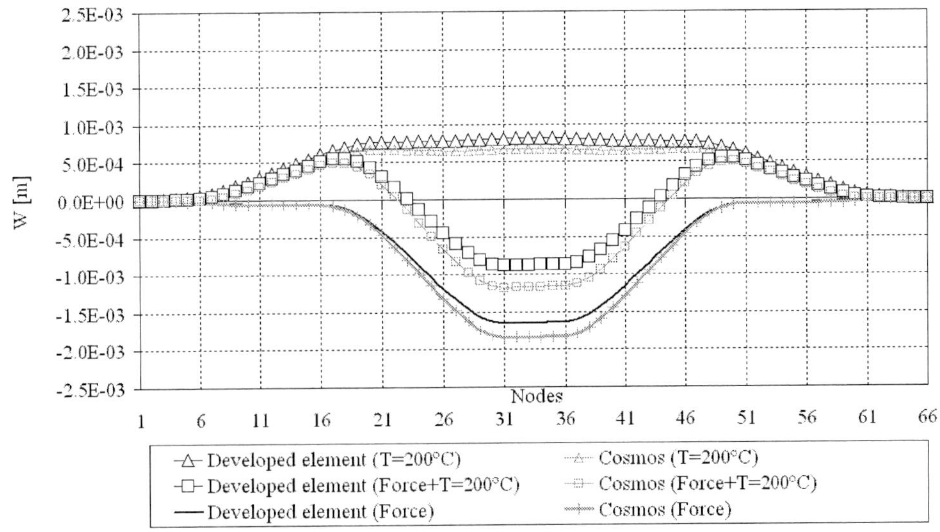

Figure 3.24. Displacement W using the pipe elbow element and COSMOS program (T=200°C).

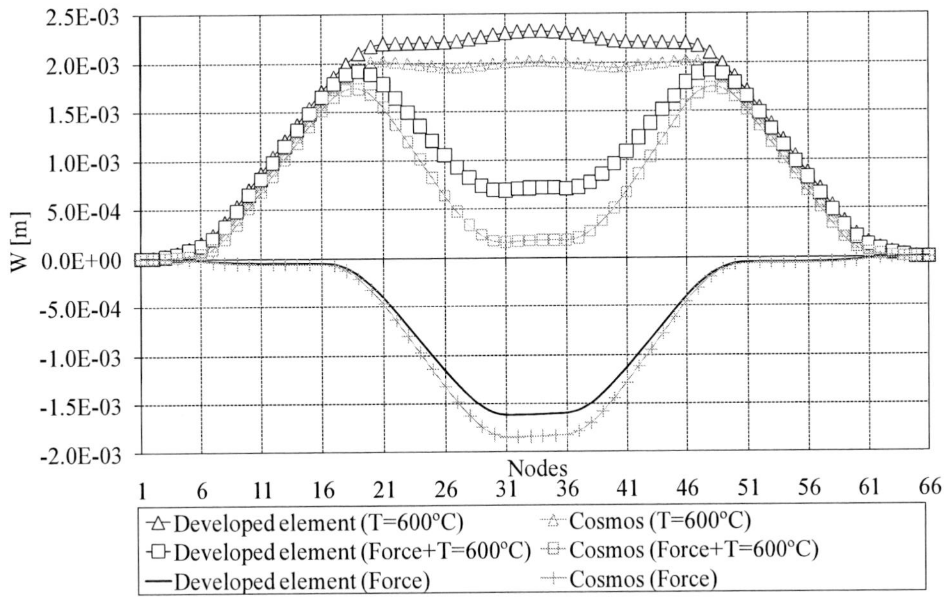

Figure 3.25. Displacement W using the pipe elbow element and COSMOS program (T=600°C).

For transversal displacement $\overline{W}$ the results also increase with the temperature, figures 3.26 and 3.27. The vertical force has influence in the results and decreases the effect of temperature.

Good agreement between the displacements results obtained with the finite element presented and corresponding data from other commercial program were observed.

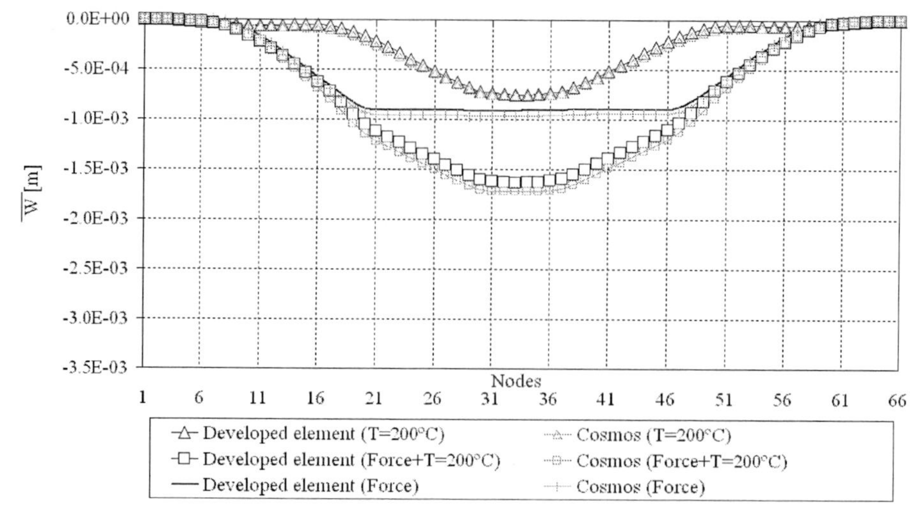

Figure 3.26. Displacement $\overline{W}$ using the pipe elbow element and COSMOS program (T=200°C).

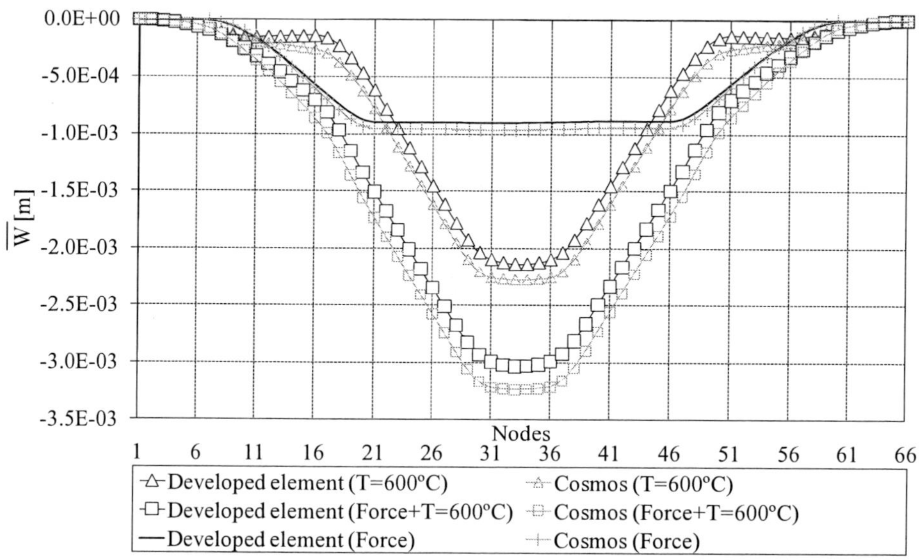

Figure 3.27. Displacement $\overline{W}$ using the pipe elbow element and COSMOS program (T=600°C).

3.5. A Spatial Pipe Elbow System Subjected to Elevated Temperatures and a Vertical Force

This case is the same as previous, but now with the spatial structure subjected to a vertical force, F=3000N at one end and the other end restrained, see figure 3.28. The structure is subjected at two uniform different temperatures, T=200°C and T=600°C.

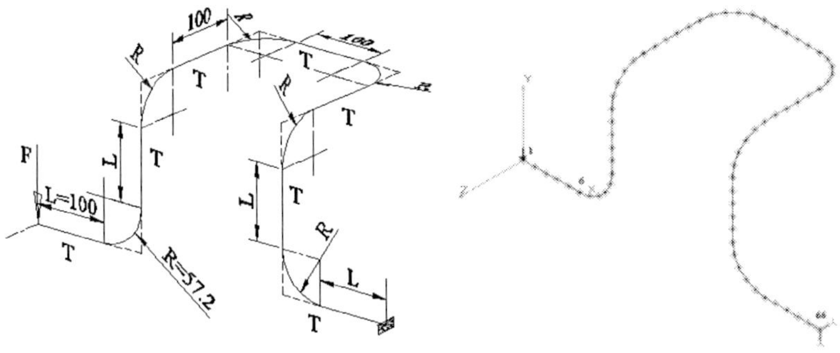

Figure 3.28. Pipe elbow system, with uniform temperature and vertical force.

The displacement field is influenced by the temperature rise, as shown in following figures, having compared these results with those corresponding to the mechanical stand alone load.

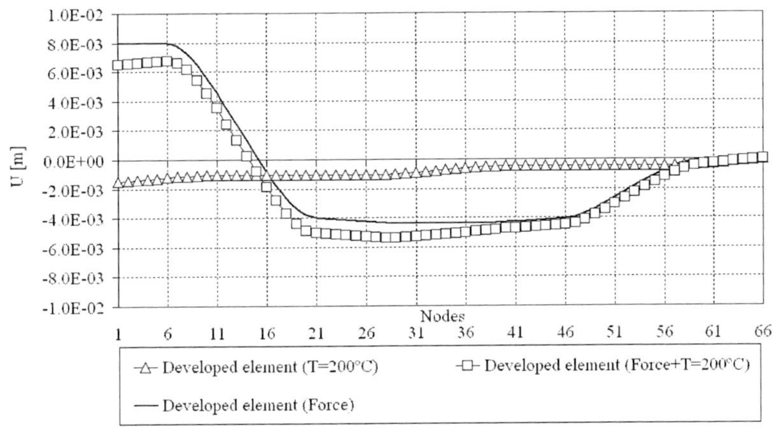

Figure 3.29. Displacement U using the pipe elbow element (T=200°C).

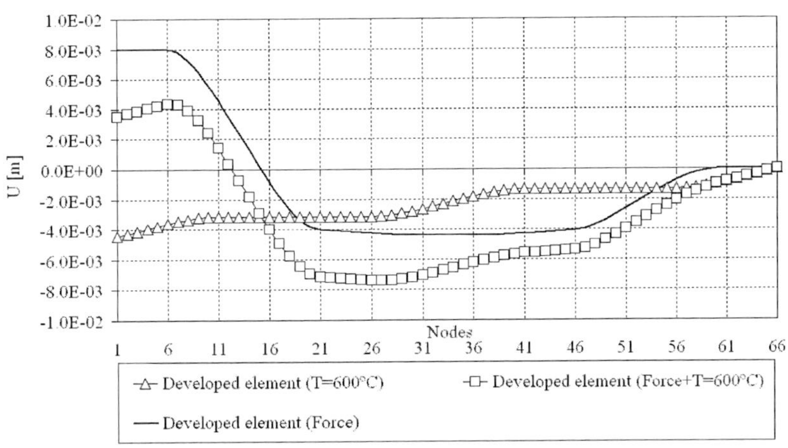

Figure 3.30. Displacement U using the pipe elbow element (T=600°C).

The displacements produced by only the vertical force present higher values. When the temperature is imposed, the longitudinal displacements decrease. The results present a constant behaviour, through all structure, when only the temperature is imposed.

The transversal displacement produced simultaneous by the vertical force and with temperature present higher values. For higher values of temperatures the displacement field increase, mainly close to the applied force.

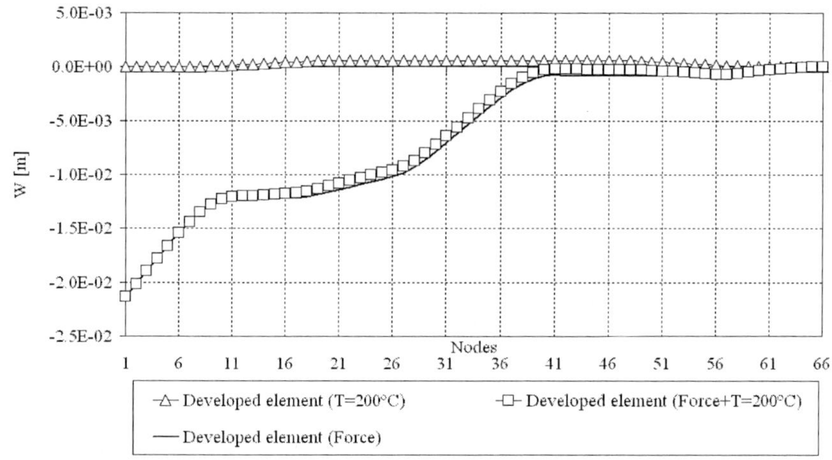

Figure 3.31. Displacement W using the pipe elbow element (T=200°C).

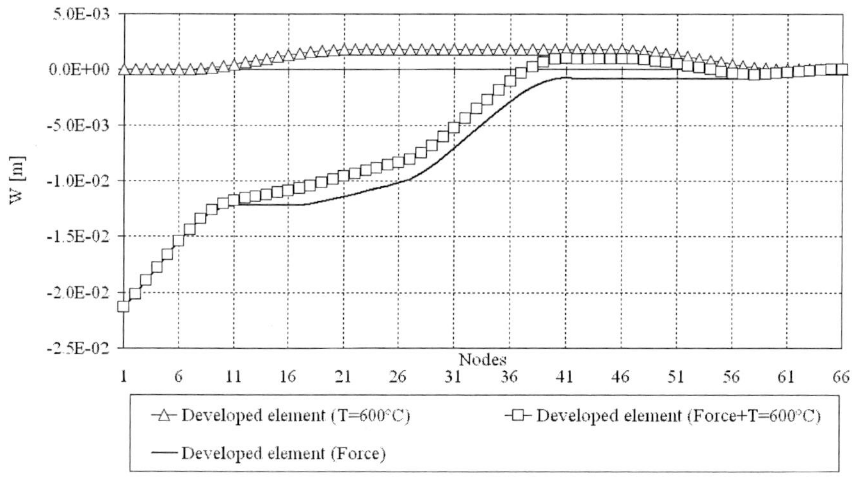

Figure 3.32. Displacement W using the pipe elbow element (T=600°C).

The transversal out-of-plane displacement produced by the vertical force has a different behaviour when the temperature is introduced. For higher values of temperatures the displacements field increase, mainly in the middle of structure.

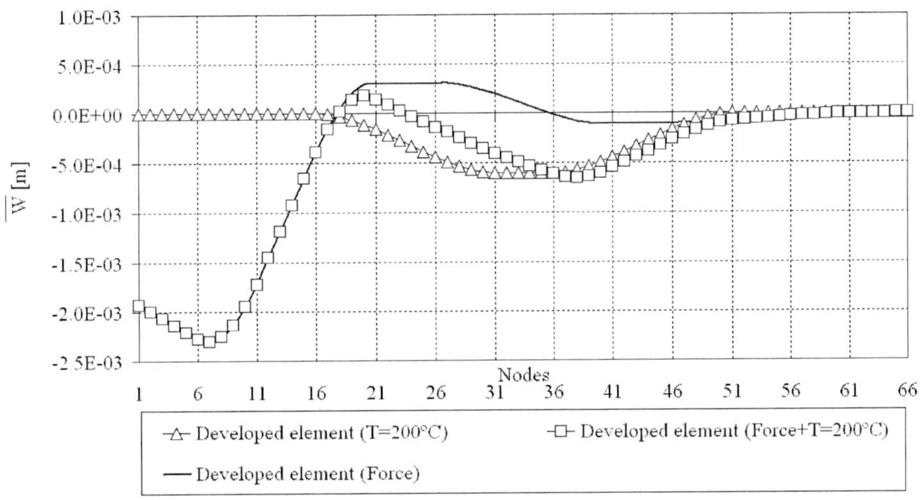

Figure 3.33. Displacement $\overline{W}$ using the pipe elbow element (T=200°C).

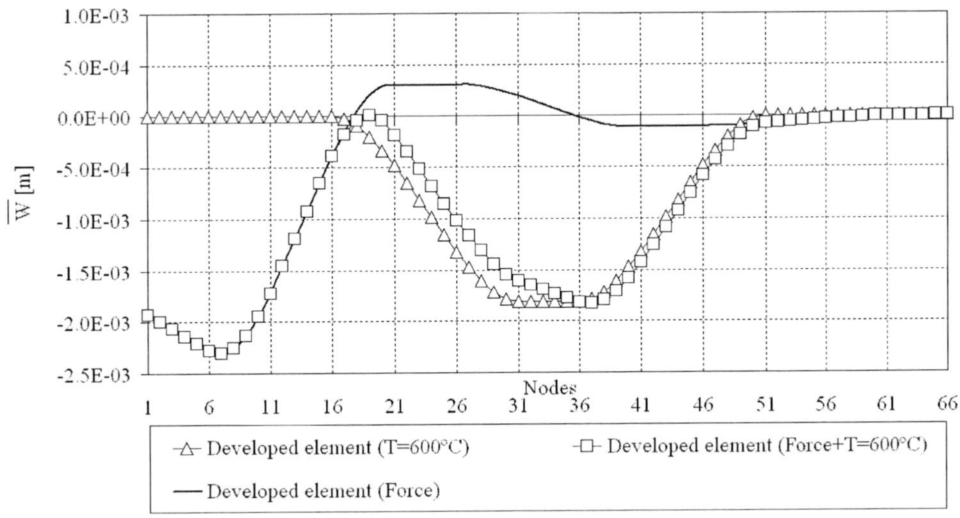

Figure 3.34. Displacement $\overline{W}$ using the pipe elbow element (T=600°C).

3.6. A Spatial Pipe Elbow System Subjected to Elevated Temperatures with Upper Length Zone Insulated and a Vertical Force

The same pipe elbow system is now subjected to different boundary conditions, a vertical force of F=3000N in free end and the other end restraint, partially subjected to uniform temperature (T=200°C or T=600°C) and partially insulated in the upper length zone AA, as shown in figure 3.35.

The displacement field increases, as can be seen in the next figures, compared with the results for previous cases.

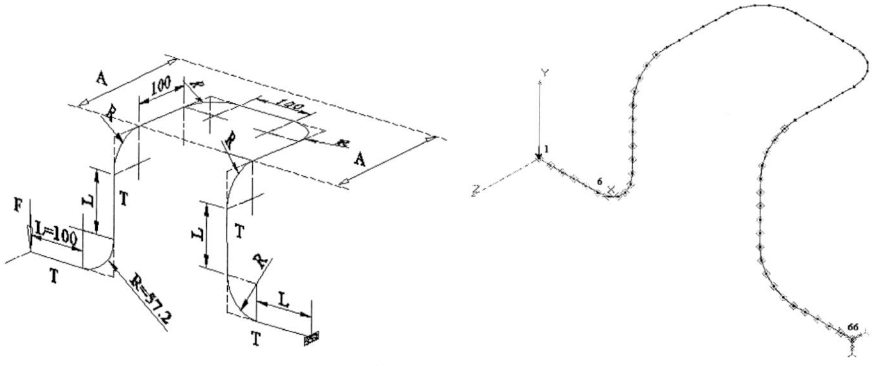

Figure 3.35. Pipe elbow system, partially subjected to uniform temperature.

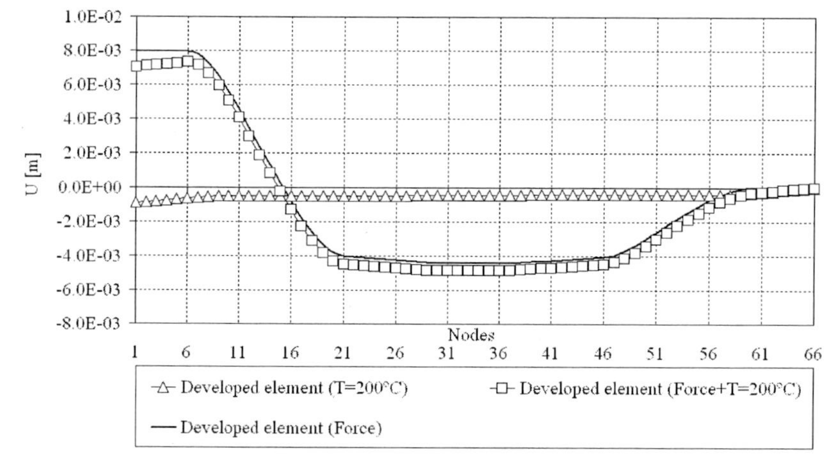

Figure 3.36. Displacement U using the pipe elbow element (T=200°C).

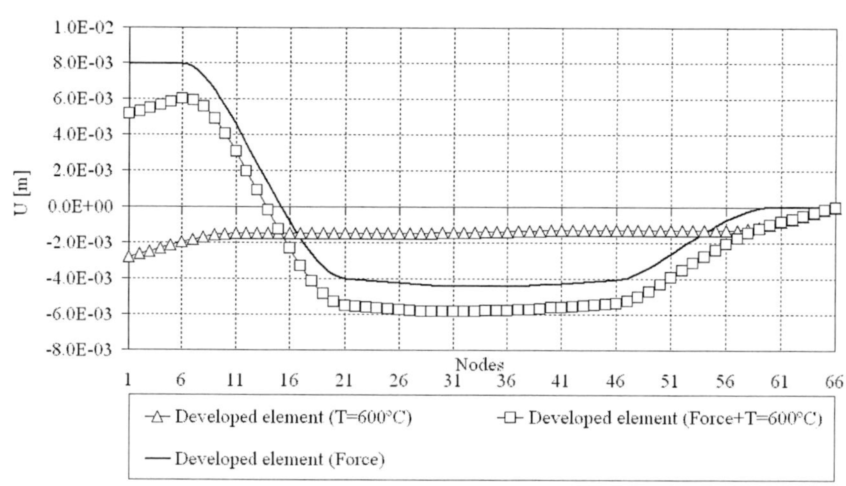

Figure 3.37. Displacement U using the pipe elbow element (T=600°C).

The longitudinal displacements U present different behaviour when increasing the temperature, figures 3.36 and 3.37.

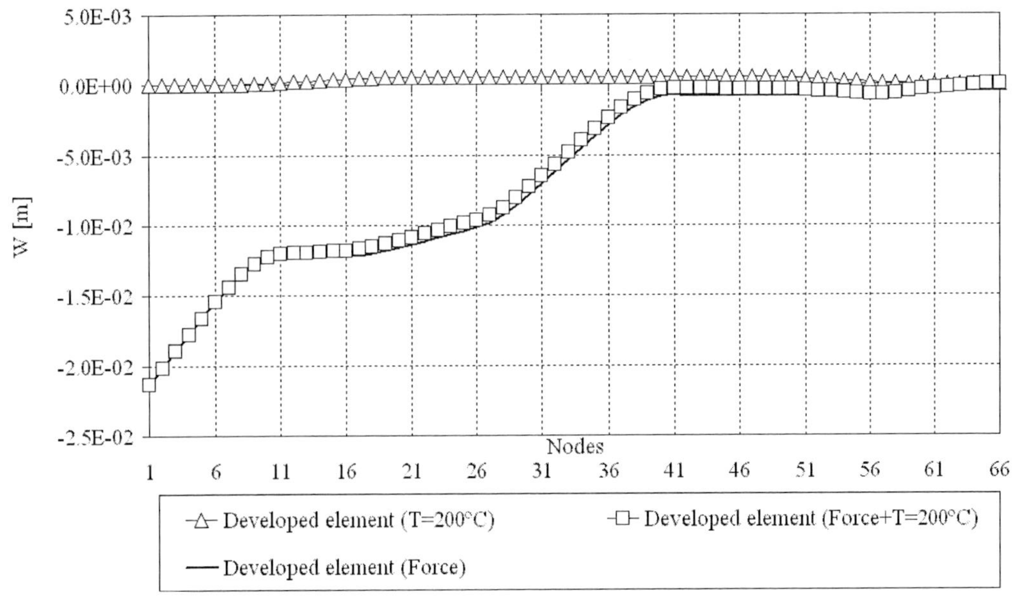

Figure 3.38. Displacement W using the pipe elbow element (T=200°C).

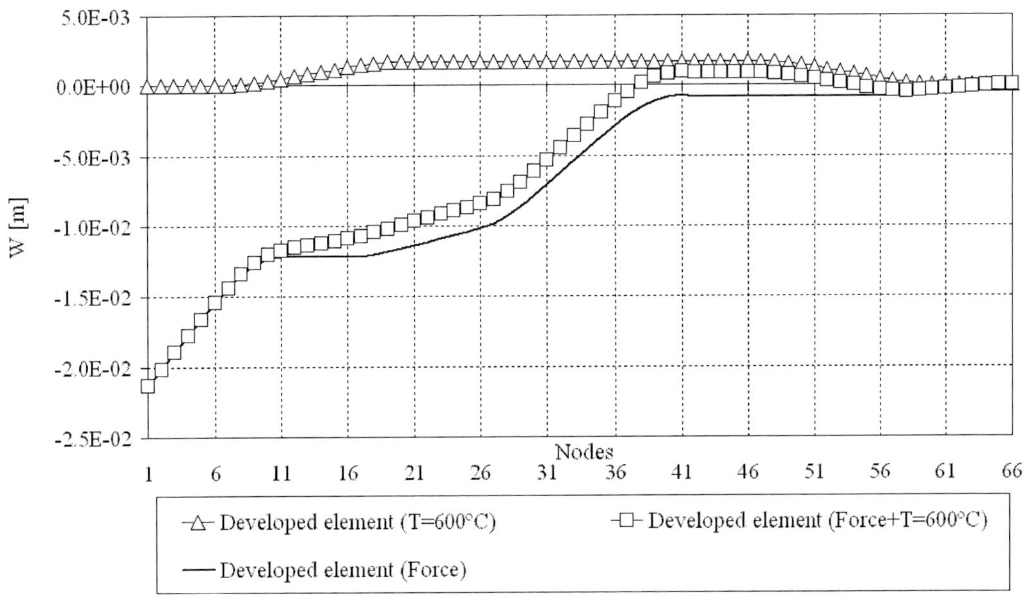

Figure 3.39. Displacement W using the pipe elbow element (T=600°C).

The transversal displacement has practically none influence due temperature evolution. The principal reason is given by the insulated upper length pipe elbow zone.

The results presented in figures 3.40 and 3.41 have an influence due temperature evolution near at insulated upper length pipe elbow zone. In this zone it is possible to verify a loosening of displacements due the temperature effect.

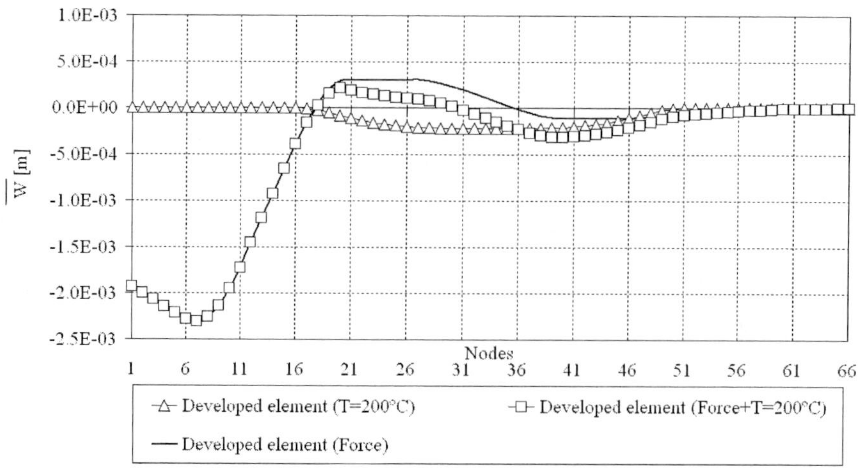

Figure 3.40. Displacement $\overline{W}$ using the pipe elbow element (T=200°C).

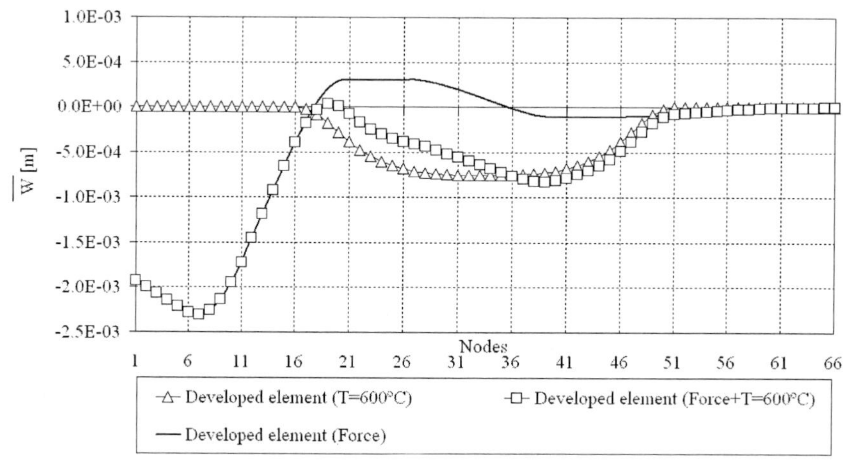

Figure 3.41. Displacement $\overline{W}$ using the pipe elbow element (T=600°C).

Conclusion

With this finite pipe elbow element it is possible to calculate the displacement field due any type of thermal and mechanical load, for in-plane and out-of-plane.

It is a simple finite element and easy to operate, avoiding expensive pre-processing mesh generation tasks for shell surface definition. The pre and post-processing operations with this pipe elbow element suggest the possibility of its use in small personal computers; in fact the finite element mesh set-up resumes to the definition of a curved line, where each node contains information for beam displacements and transverse section distortion.

The present method is a simple and economic tool for the determination of the stress field in pipe elements using a special finite element formulation. The obtained values of the stress distribution along transverse sections showed a good agreement with corresponding results from other analyses. The presented solution has performed well even with coarse element meshes.

Several case studies presented were compared and discussed with results reported by other authors. Also to validate the element accuracy, numerical results using alternative commercial software were presented. Also the results for structural displacements resulting from elevated temperatures and mechanical actions on tubular structures have been presented.

The purpose of this work is to provide an easy and an alternative formulation when compared with a complex finite shell, solid or beam element analysis for the same application.

About Author's

E.M.M. Fonseca received the Ph.D. degree in Mechanical Engineering from University of Porto, Portugal, 2003. She is Assistant Professor of the Applied Mechanics Department at the Polytechnic Institute of Bragança, Bragança, Portugal, since 1995. She was Postdoctor of University of Aveiro through a research fellow of FCT/Science and Technology Foundation. Researcher of CENUME-IDMEC, Unit for Numerical Methods in Mechanics and Structural Engineering since 1996. Her research interests include the solid mechanics, computational mechanics, piping structures, fire and biomechanics. She was member of the International and National Scientific Committees in different conferences. Referee for different journals. She belongs to the Editorial Board of the International Journal of Safety and Security Engineering and Journal of Experimental Mechanics.

F.J.M.Q. Melo Received the graduation in Mechanical Engineering, FEUP, Faculty of Engineering, University of Porto, Portugal, in 1975; MSc, Mechanical Engineering in 1984, FEUP and PhD in Mechanical Engineering, also in FEUP, 1988. Actual activity and position: Associate Professor in the Department of Mechanical Engineering, University of Aveiro, Campus of Santiago 3810-Aveiro Portugal, teaching classes of Introduction to Mechanical Design, Advanced Structural design (as a PhD complement program class) and Topics of MSc Theses. Researcher in TEMA, FCT (Portuguese Foundation for Science and Technology) registered research unity of Mechanical Technology and Automation; area of Advanced Mechanical Engineering. Areas of research: Numerical modelling of structural dynamics (structural vibrations, Hopkinson Pressure bar models for simulation of materials at high strain rates), experience in automotive body design and construction, aeronautics and aerospace construction; Simulation of material behaviour for application in sheet metal forming techniques (stamping and incremental sheet forming); Coordination of design solutions in the structural assessment of heavy welded construction steel frames for sheet metal forming mechanical presses.

L.R.Madureira. received the Ph.D. degree in Mechanical Engineering from University of Porto, Portugal, 1996. She is Professor of the Department of Mechanical Engineering of the Faculty of Engineering of University of Porto, since 1984, Auxiliar Professor since 1996. Researcher of CENUME-IDMEC, Unit for Numerical Methods in Mechanics and Structural Engineering since 1996. Her research interests include the solid mechanics, computational mechanics, piping structures, mathematics. Referee for Int Jour of Pressure Vessels and Piping and Int Jour of Structural Integrity.

REFERENCES

Bathe K. J., Almeida C. A., A simple and effective pipe elbow element–Pressure stiffening effects. *J. Appl. Mechanics*, 1982, Vol.49, 914-916.

CEN ENV 1993-1-2, Eurocode 3: *Design of Steel Structures - Part 1.2: General Rules - Structural Fire Design*, 1995.

Cheng D. H., Thailer M. J., On the bending of curved circular tubes. ASME, *J. Eng. Industry*, 1970, Vol.92(1) B, 62-66.

Flügge W., Thin elastic shells. 2nd ed., Berlin: Springer, 1973.

Fonseca E. M. M., Melo F. J. M. Q., Numerical analysis of curved pipes submitted to in-plane loading conditions. *Thin-Walled Structures*, 2010, Vol.48, 103-109.

Fonseca E. M. M., Melo F. J. M. Q., Oliveira C. A. M., Determination of Flexibility Factors on Curved Pipes with end Restraints Using a Semi-Analytic Formulation, *International Journal of Pressure Vessels and Piping*, 2002, Vol.79(12), 829-840.

Fonseca E. M. M., Melo F. J. M. Q., Oliveira C. A. M., Numerical analysis of piping elbows for in-plane bending and internal pressure. *Thin-Walled Structures*, 2006, Vol.44, 393-398.

Fonseca E. M. M., Melo F. J. M. Q., Oliveira C. A. M., The thermal and mechanical behaviour of structural steel piping systems. *International Journal of Pressure Vessels and Piping*, 2005; Vol.82(2), 145-153.

Hose D., Kitching R., Behaviour of GRP smooth pipe bends with tangent pipes under flexure or pressure loads: a comparison of analyses by conventional and finite element techniques", *Int. J. Mech. Sci.,* 1993, Vol.35(7), 549-575.

Kitching R., Hughes J. F., Pad-reinforced cylindrical shell loaded radially through a plate bracket attachment, *The Journal of Strain Analysis for Engineering Design Professional Engineering Publishing*, 1980, Vol.15(1), 1-14.

Kitching R., Smooth and mitred pipe bends. Gill S.S. (Ed.), *The stress analysis of pressure vessels and pressure vessels components*, Chapter 7, Pergamon Press: Oxford, 1970.

Love A. E. H.. *A treatise on the mathematical theory of elasticity*. New York: Dover, 1944.

Madureira L., Melo F. Q., A Hybrid Formulation in the Stress Analysis of Curved Pipes. *Engineering Computations*, 2000, Vol.17(8), 970-980.

Madureira L., Melo F. Q., Stress analysis of curved pipes with hybrid formulation, *Pressure Vessels and Piping* 2004, Vol.8(3), 243-249.

Melo F. J. M. Q., Castro P. M. S. T., A reduced integration Mindlin beam element for linear elastic stress analysis of curved pipes under generalized in-plane loading. *Computers and Structures*, 1992, Vol.43(4), 787-794.

Melo F. J. M. Q., Castro P. M. S. T., The linear elastic stress analysis of curved pipes under generalized loads using a reduced integration finite ring element. *Journal of Strain Analysis*, 1997, Vol.32(1), 47-59.

Ohtsubo H., Watanabe O., Stress analysis of pipe bends by ring elements. *Journal of Pressure Vesssel Technology*, 1978, Vol.100, 112-121.

Oñate E., E, *Cálculo de Estruturas por el Método de Elementos Finitos,* CIMNE, Barcelona, 1992.

Öry H., Wilczek E., Stress and stiffness calculation of thin-walled curved pipes with realistic boundary conditions being loaded in the plane of curvature. *International Journal of Pressure Vessels and Piping* 1983, Vol.12, 167-189.

Rhodes, *Thin-Walled Structures*, 1997, Vol.28(3-4), 201-212.

Thomson G., The influence of end constraints on pipe bends. PhD Thesis, University of Strathclyde, Scotland, UK, 1980.

Vigness I., Elastic properties of curved tubes. *ASME*, 65, 105-120, 1943.

von Karman Theodore, Über die formanderung dunnwaindiger rohre insbesondere federnder ausgleichrohre. *Zeits V.D.I., Band* 55, ss. 889-1895, 1911.

Wilczek E., Statische berechnung eines rohrkrümmers mit realen randbedingungen. Ph.D. thesis, Institut für Leichtbau, Aachen, 1984.

In: Toroidal Shells
Editor: Bohua Sun
ISBN: 978-1-61942-247-6

Chapter 5

VIBRATION OF TOROIDAL SHELLS AND CURVED TUBES*

Xiaohong Wang†
Department of Civil Engineering, Shantou University, Shantou, China

1. INTRODUCTION

Vibration problems in structures of various shapes and configurations have been extensively studied in the past hundred years or so where many pioneers, Rayleigh and Love in particular, had made considerable contributions in shell structures. Since then, the importance of shell theories can be recognized by the numerous reference books and reviews during the past few decades. In particular, the two distinguished monographs on shell vibration analysis from Leissa (1973) and Soedel (2004) enriched the theories that have broad applications still applicable even today.

Research on toroidal shells began a half century ago, the early work by Clark (1950) establishing the theory of thin elastic toroidal shells. Later in a series of studies, Sanders and Liepins (1963) investigated the toroidal membrane under uniform pressure. Their approach was improved by Rossettos and Sanders (1965) who used both the nonlinear membrane theory and the bending theory. Liepins (1965) studied the free vibrations of prestressed toroidal shells. His research was based on the linearized shell theory and applied Fourier series expansions in the circumferential direction and finite difference approximations in the meridional direction. Balderes and Armenakas (1973) first investigated the free vibration of ring-stiffened isotropic toroidal shells using the Love-Reissner shell theory. Using a toroidal shell with holes as an example, Gavelya et al (1975) calculated the Green's matrices used to determine the stress-strain state of shells. Kosawada et al (1985) presented the free vibration of thin and thick toroidal shells with circular cross-section. Yamada et al (1989) presented the solution to the free vibration problem of a toroidal shell with elliptical cross-section using the

---

* This paper is based on the author's master and postdoctoral research in toroidal shell structures. The author has been working in this field since 2002 and the research work has been published in different conferences and scientific journals.

† e-mail: wangxh@stu.edu.cn.

transfer matrix approach. Leung and Kwok (1994) studied a 90-degree-bend curved pipe based on Donnell-Mushtari-Vlasov thin shell theory, Fourier's series and Galerkin's method. Redekop and his co-workers conducted extensive research work on the static, dynamic, vibration, and buckling characteristics of curved tubes and toroidal shells [Redekop(1992), (1994a), (1994b), (1994c), (1995), (1997), (1999), (2000), (2004a), (2004b), (2005), (2006)] [Redekop and Xu(1999), (2000)] [Wang and Redekop(2005)] [Jiang et al (2005), Redekop and Muhammad (2003), Xu and Redekop (2006)][Zhan and Redekop (2008a), (2008b)]. Huang et al (1997) used the linear elastic Sanders shell theory to solve the problem of the free vibration of a curved pipe with rigid diaphragm end supports. Redekop and Xu (1999) studied the vibration of toroidal panels using the differential quadrature method (DQM). Ming et al (2002) experimentally developed the free vibrations of a circular toroidal shell under different boundary conditions. Li and Cook (2002) investigated fiber overwrapped toroidal vessels using membrane theory. Jiang and Redekop (2003) presented the static and vibration analysis of orthotropic toroidal shells of variable thickness by the DQM. Wang et al (2006) provided the extensive results for frequencies of vibrations of thin and thick curved pipes and toroidal shells determined by either analytical or numerical methods.

In the last two decades, there has been tremendous progress in the applications and analysis of toroidal shells. Several recent papers have presented reviews on toroidal shells, each giving an important partial view of the subject. In 1999, Ren et al presented a review of studies on toroidal shells and curved tubes that covered mainly theoretical work done in China and Germany. Ruggiero et al (2003) presented a review of experimental and theoretical work on toroidal satellite (gossamer) structures. Later Ruggiero and Inman further discussed the development trends in design, analysis, experimentation and control in gossamer spacecrafts [Ruggiero and Inman (2006)].

2. Thin Shell Theory

The thin shell theory is firmly established in the literature and is used extensively in analytical solutions and numerical analysis [Novozhilov(1959), Flügge(1967), Kraus(1967),Timoshenko and Woinowsky-Krieger(1959)]. The most common shell theories are those based on linear elastic concepts. The first-order Sanders-Budiansky [Sanders(1959), Budiansky(1968)] linear thin shell theory applied for toroidal shells is discussed in this section. This theory is formulated based on the Love-Kirchoff assumptions that normal to the shell middle surface remains normal to it during its vibration and unstretched in length, and it is considered one of the most accurate first-order theories [Librescu(1975), Markus(1988)].

2.1. Background

Toroidal shells shown in Figure 2.1 are one of the typical shells of revolution and they have been integrated into many engineering applications, such as piping systems, pressure vessels technology, satellite antenna support structures, inflatable space structures, rocket fuel tanks, and protective devices for nuclear waste containers.

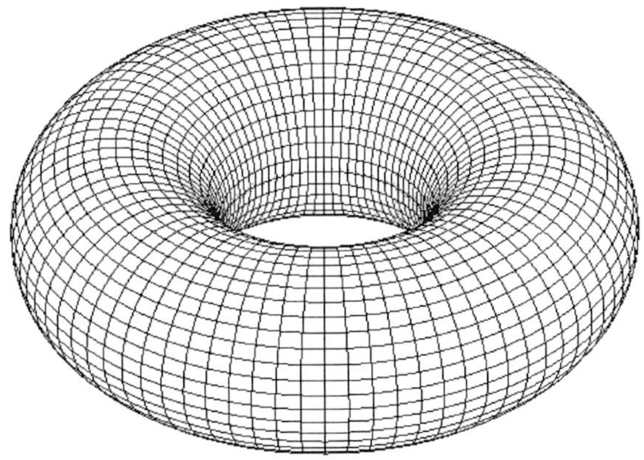

Figure 2.1. Complete circular toroidal shell.

There are a vast number of published literatures existing in free vibration analysis for thin shells of revolution. The early work on free vibration analysis for shells of revolution can be found in the study of Bacon and Bert(1967). In their study, they analyzed free vibrations of shells of revolution numerically using Love's first-approximation shell theory with considering the effect of transverse-shear deformation, but the application was limited to the simple case of cylindrical shells and the conical-frustum shells. In Leissa's monographs (1973) on shell vibration analysis, he summarized the approximately 1000 relevant publications world-wide through the 1960s, and most of these dealt with shells of revolution with applications intended for circular cylindrical, conical, spherical and paraboloidal shells.

Later, Sen and Grould (1974) presented a displacement finite element formulation for the vibration analysis of shells of revolution and solved the eigenvalue problem by employing the kinematic condensation technique. In recent literature on the vibration of shells of revolution, Tan (1998) developed a substructuring technique to analyze nature frequencies and considered applications to thin cylindrical, spherical, and hyperbolical shells. Noor and Peters (2002) investigated an efficient computational strategy to analyze the stress and free vibration of laminated anisotropic shells of revolution based on Sanders-Budiansky shell theory, and the numerical results solved using FEM. More recently, Redekop (2004a) used the DQM to study bodies of revolution of arbitrary thickness, and considered cylindrical and hemispherical applications. To consider the important position of toroidal shells in engineering applications, it is desirable and advantageous to develop the theory for general shells of revolution not only toroidal shells can be analyzed but also other shells such as the cylindrical and spherical. Therefore, in the study of this section, theory is first derived for shells of revolution, and then applied for toroidal shells.

2.2. Geometry and Boundary Conditions

The mid-surface of an arbitrary thin shell can be described by a radius vector R = R (q_1, q_2), where q_1, q_2 form an orthogonal coordinate system [Wang(2004)]. α_1, α_2, and κ_1, κ_2 represent the Lamé parameters of the shell and the curvatures, respectively. The radius vector of a shell of revolution takes the form R = $\bar{r}\sin\phi\,\mathrm{i} + \bar{r}\cos\phi\,\mathrm{j}$ + zk, where $q_1 = \phi$ is the circumferential angle, $\bar{r} = \bar{r}(q_2)$, $z = z(q_2)$, and i, j, k are the Cartesian unit vectors [Dym(1974)], respectively.

For a specific shell of revolution, the geometric parameters may readily be derived from the radius vector [Soedel(2004)]. As an example, a toroidal shell with circular cross-section shown in Figure 2.2, its corresponding geometric parameters are given by

$$q_1 = \phi;\ q_2 = \theta;\ \alpha_1 = R + r\cos\theta;\ \alpha_2 = r;\ \kappa_1 = \frac{\cos\theta}{R + r\cos\theta};\ \kappa_2 = \frac{1}{r} \tag{2.1}$$

where ϕ is the circumferential angle, θ is the meridional angle, measured clockwise from the positive horizontal, R is the bend radius, r is the radius of the cross-section, and $z = z(\theta) = r\cos\theta$.

The meridian is a closed curve in a complete toroidal shell, and boundary conditions are not considered for this geometry of a toroidal shell. This proposed method can be extended to shells of revolution considering incomplete meridian specified with appropriate boundary conditions at the ends of the meridians for such shells.

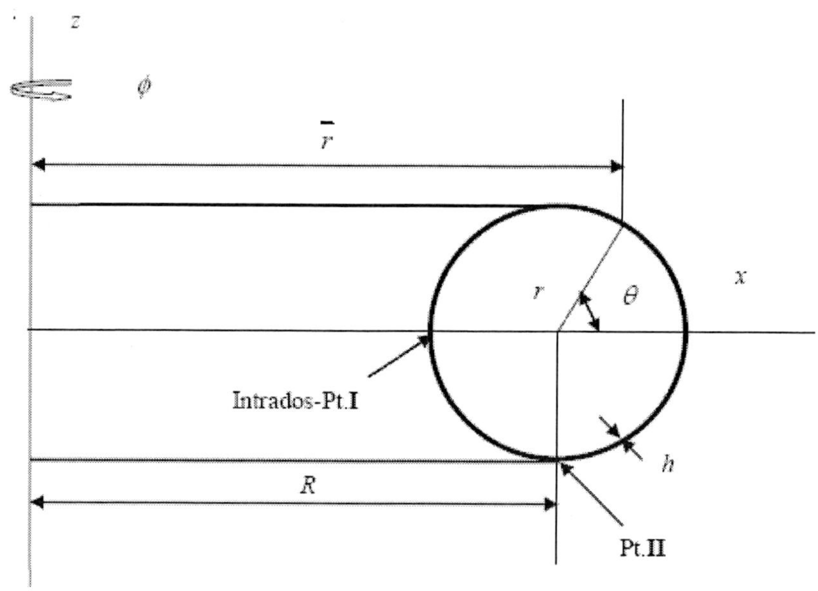

Figure 2.2. Cross-sectional geometry of circular toroidal shell; point I is the intrados, point II is the lower crown.

2.3. Thin Shell Theory

In this section, Sanders-Budiansky shell theory [Sanders(1959), Budiansky(1968)] is used to determine the vibration characteristics of the shell of revolution associate with the D'Alembert principle and isotropic constitutive relations. Transverse shear effects or other higher order effects are not considered in this analysis. And inertias arising from in-plane and transverse motion are considered, but rotary inertia effects are neglected.

2.3.1. Equilibrium Equations

Adopting notations $(\ )^{\bullet} = \dfrac{\partial}{\partial \phi}(\ )$ and $(\ )' = \dfrac{\partial}{\partial \theta}(\ )$, for a general shell of revolution on vibration analysis, in terms of the stress and moment resultants, the equations of motion in this theory can be expressed as [Redekop(2004b), Wang(2004)]:

$$\begin{aligned} &a_1 N_\phi^{\bullet} + a_2 N_\phi + a_3 N_\theta + a_4 N_\phi' + a_5 N_{\phi\theta} + a_6 M_\phi^{\bullet} + a_7 M_\phi + a_8 M_\theta \\ &+ a_9 M_{\phi\theta}' + a_{10} M_{\phi\theta} + \alpha_1 \alpha_2 \rho h \ddot{u}_1 = 0 \\[1em] &a_{11} N_\phi + a_{12} N_\theta' + a_{13} N_\theta + a_{14} N_\phi' + a_{15} N_{\phi\theta} + a_{16} M_\phi + a_{17} M_\theta' + a_{18} M_\theta \\ &+ a_{19} M_{\phi\theta}^{\bullet} + a_{20} M_{\phi\theta} + \alpha_1 \alpha_2 \rho h\, \ddot{u}_2 = 0 \\[1em] &a_{21} N_\phi + a_{22} N_\theta + a_{23} M_\phi^{\bullet\bullet} + a_{24} M_\phi^{\bullet} + a_{25} M_\phi' + a_{26} M_\phi + a_{27} M_\theta^{\bullet} + a_{28} M_\theta'' \\ &+ a_{29} M_\theta' + a_{30} M_\theta + a_{31} M_{\theta\phi}^{\bullet} + a_{32} M_{\theta\phi}'^{\bullet} + a_{33} M_\theta^{\bullet}{}' + a_{34} M_{\phi\theta} + \alpha_1 \alpha_2 \rho h \ddot{u}_3 = 0 \end{aligned} \quad (2.2)$$

where ρ is the mass density, h is the shell thickness, $\ddot{u}_1, \ddot{u}_2, \ddot{u}_3$ are the acceleration components, and the a_i, $i = 1,\ldots,34$ are known functions of the Lamé parameters and curvatures.

The constitutive relations are taken as the followings for an elastic and isotropic shell with Young's moduli E, Poisson ratios ν [Sanders(1959)]:

$$\begin{aligned} &N_\phi = K(\varepsilon_\phi + \nu\varepsilon_\theta),\ N_\theta = K(\varepsilon_\theta + \nu\varepsilon_\phi),\ N_{\phi\theta} = \frac{K(1-\nu)}{2}\varepsilon_{\phi\theta} \\ &M_\phi = D(\kappa_\phi + \nu\kappa_\theta),\ M_\theta = D(\kappa_\theta + \nu\kappa_\phi),\ M_{\phi\theta} = D(1-\nu)\kappa_{\phi\theta} \end{aligned} \quad (2.3)$$

where K and D defined as:

$$K = \frac{Eh}{1-\nu^2},\ D = \frac{Eh^3}{12(1-\nu^2)}$$

2.3.2. Strain-Displacement Relationships

In ϕ and θ directions, $\varepsilon_\phi, \varepsilon_\theta, \varepsilon_{\phi\theta}$ are the strain components, and $\kappa_\phi, \kappa_\theta, \kappa_{\phi\theta}$ are the bending strains. Based on the Sanders-Budiansky theory, the kinematic relations are given as [Wang(2004)]

$$\varepsilon_\phi = c_1 u_1^{\bullet} + c_2 u_2 + c_3 u_3$$

$$\varepsilon_\theta = c_4 u_1 + c_5 u_2' + c_6 u_3$$

$$\varepsilon_{\phi\theta} = c_7 u_1' + c_8 u_1 + c_9 \dot{u}_2 + c_{10} u_2$$

$$\kappa_{\phi} = c_{11} \dot{u}_1 + c_{12} u_1 + c_{13} u_2 + c_{14} \ddot{u}_3 + c_{15} \dot{u}_3 + c_{16} u_3' \tag{2.4}$$

$$\kappa_{\theta} = c_{17} u_1 + c_{18} u_2' + c_{19} u_2 + c_{20} \dot{u}_3 + c_{21} u_3'' + c_{22} u_3$$

$$\kappa_{\phi\theta} = c_{23} u_1' + c_{24} u_1 + c_{25} \dot{u}_2 + c_{26} u_2 + c_{27} \dot{u}_3 + c_{28} u_3' + c_{29} \dot{u}_3'$$

where the c_i, $i = 1,\ldots, 29$ are known functions of the Lamé parameters and curvatures.

The displacement components of the mth circumferential harmonic are represented using a semi-analytical approach.

$$u_1 = u(\theta)\cos m\phi \sin \omega t$$

$$u_2 = v(\theta)\sin m\phi \sin \omega t \tag{2.5}$$

$$u_3 = w(\theta)\sin m\phi \sin \omega t$$

where the u, v, w represent the displacement functions for the mth harmonic mode dependent on θ only, ω is the natural frequency, and t is time.

Substituting the displacement components (Eq. 2.5) into the kinematic relations (Eq. (2.4)) together with the constitutive relations (Eq. 2.3) and the equations of motion (Eq. 2.2), then one can obtain the three governing equations as [Wang(2004)]

$$\begin{aligned}
&e_1 u'' + e_2 u' + (e_3 + m^2 e_4) u + m e_5 v' + m e_6 v + m e_7 w'' \\
&+ m e_8 w' + (m^3 e_9 + m e_{10}) w = \Omega u \\[2ex]
&m e_{11} u' + m e_{12} u + e_{13} v'' + m e_{14} v' + (e_{15} + m^2 e_{16}) v + m e_{17} w''' \\
&+ m e_{18} w'' + (e_{19} + m^2 e_{20}) w' + (e_{21} + m^2 e_{22}) w = \Omega v \\[2ex]
&m e_{23} u'' + m e_{24} u' + (m^3 e_{25} + m e_{26}) u + e_{27} v''' + e_{28} v'' \\
&+ (e_{29} + m^2 e_{30}) v' + (m^2 e_{31} + e_{32}) v + e_{33} w'''' + e_{34} w''' \\
&+ (e_{35} + m^2 e_{36}) w'' + (e_{37} + m^2 e_{38}) w' + (e_{39} + m^2 e_{40} + m^4 e_{41}) w = \Omega w
\end{aligned} \tag{2.6}$$

where the e_i, $i = 1,\ldots,41$ are known functions of the Lamé parameters, curvatures, material properties, and $\Omega = \rho h \omega^2$. The displacement components together with the natural frequency parameter of the mth harmonic as the unknowns are included in the above three differential equations (Eq. 2.6).

The axi-symmetric circumferential harmonic, i.e. $m = 0$ is a special case, in which the displacement u_ϕ, the resultants $N_{\phi\theta}$, $M_{\phi\theta}$ and their derivatives with respect to ϕ are all set to zero.

2.4. Differential Quadrature Method

There are many numerical methods available for shell vibration analysis, such as boundary element method (BEM), finite difference method (FDM), FEM, DQM, Rayleigh-Ritz method (RRM), etc. The DQM, which is recognized as a powerful numerical method for the study of vibration analysis of the structures, is used to carry out the numerical results for nature frequencies of toroidal shells in this study [Bert and Malik(1996a), (1996b)].

The DQM that transforms governing equations of differential form to matrix form by using weighting matrices, is a computationally efficient method for solving differential equations. There are two steps involved in this approach. A grid of sampling points covering the domain and including the boundary is required to define at the first step. In the current study, the problem has been reduced to one-dimension mathematically, which only requires a grid along an arbitrary meridional line. Therefore the three displacements components associated with frequency at each sampling points of the grid comprise the unknowns of the problem.

The second step of the DQM approach is involved with the replacement of all derivatives with a series of terms that contain the product of the displacement functions at the sampling points and the weighting coefficients. This step allows the problem transformed from one of differential equations to one of linear equations. At the sampling point x_i, the r-th derivative of a generic function of a single variable $f(x)$ is expressed as

$$\left.\frac{d^r f(x)}{dx^r}\right|_{x_i} = \sum_{h=1}^{M} A_{ih}^{(r)} f(x_h) \qquad (2.7)$$

where M is the number of sampling points in the x direction, $A_{ih}^{(r)}$ are the weighting coefficients of the r-th order derivative in the x direction for the i-th sampling point and $f(x_h)$ is the value of $f(x)$ at the sampling point position x_h.

In the DQM, with the aid of selected trial functions, the weighting coefficients are determined a-prior for a preselected grid. Based on the current study, since the meridional direction is the closed form in geometry, it is convenient to use a Fourier basis for the weighting coefficients with equally spaced points. The explicit formulas of the weighting coefficients $A_{ih}^{(r)}$ for this problem can be found in Shu (2000).

Using the quadrature rule of Eq. (2.7) for the derivatives of the governing differential equations (Eq. (2.6) leads to a set of simultaneous linear equations of the form

$$[\mathrm{K}](U) = \lambda\, [\mathrm{M}]\, (U) \qquad (2.8)$$

where [K], [M] are the known ‘stiffness’ and ‘mass’ matrices, the unknowns (U) are the values of the displacement functions at the sampling points, and λ is the eigenvalue related to ω. Then the standard matrix eigenvalue extraction routines can be used to solve the Eq. (2.8).

2.5. Validation and Results

The accuracy of this section study can be validated by comparing the DQM results obtained from this section with the results presented previously by Balderes and Armenakas (1973) using the Runge-Kutta method (RKM). The FEM results based on computer software ADINA (2003) is to provide the alternate solutions in this validation as well [Wang(2004)].

The geometry and material properties of a toroidal shell in this validation is described in Table 2-1. And the verification results are demonstrated in Table 2-2.

Table 2.1. Geometry and material properties of a toroidal shell

Properties	Values
Bend radius: R (m)	1.0
Radius of cross-section: r (m)	0.2
Thickness: h (m)	0.004
Elastic modulus: E (N/m^2)	0.207e12
Poisson’s ratio: v	0.3
Density: ρ (kg/m^3)	7800

Table 2.2. Method validation-comparison of ω (Hz) with FEM and RKM [Balderes and Armenakas (1973); Wang (2004)]

Method	ω_1	ω_2	ω_3	ω_4	ω_5	ω_6
m	2	2	0	3	3	4
DQM	487.3	491.6	631.1	985.0	996.9	1400.4
FEM	488.0	492.0	631.7	986.2	998.2	1402.6
RKM	489.0	493.0	630.9	989.6	991.0	1408.4

The DQM and FEM results in Table 2 presented, show good agreement with the RKM within 0.6%. The results between DQM and FEM demonstrated excellent agreement with maximum differences less than 0.2%. The good agreement for the results based on the three different methods indicates the accuracy of the current method and also indicates the near equivalence of the underlying shell theories for the three methods.

3. Shear Deformation Theory

As discussed previously, the theory and applications for toroidal shells are all limited on thin-walled structures. The wall thickness is a critical factor to affect the vibration analysis of shell structures. With more applications of toroidal shells in engineering practice, such as piping technology, pressure vessels, space vehicle, liquid storage structures and water tanks, they cannot be considered as thin-walled structures. For such applications, the corresponding wall thickness has to be moderately thick or thick and the analysis theory is required using the shear deformation theory (SDT) or elasticity theory (ELT). It is imperative and desirable to establish effective theories appropriate for the vibration analysis for toroidal shells based on their various thicknesses. The present section extends the vibration analysis of toroidal shells using thin shell theory to the analysis for moderately-thick and thick toroidal shells. In this section, the vibration analysis using SDT which covers the moderately thick toroidal shells is discussed. The vibration analysis for thick toroidal shells using ELT will be discussed in the next section.

3.1. Background

The research work for the analysis of toroidal shell vibration based on the classical (Love-Kirchoff) shell theory has been the subject of numerous investigations. This theory is only applied for the relatively thin shells. The early study considering the effects of shear deformation for shell vibration can be found in Silva (1964). In his study, he investigated the effect of transverse shear deformation on the bending of elliptic toroidal shells and demonstrated the transverse shear deformation becomes important as the wall thickness increases. There is further research on toroidal shell vibration using the shear deformation theory applicable for thicker shells. Kosawada et al (1986) developed a solution for thick toroidal shell vibrations and their study however included an awkward series solution in the thickness direction. In recent studies, Artioli et al (2005) and Artioli and Viola (2006) presented a shell of revolution using the shear deformation theory, but the application of their studies was only intended for paraboloidal shells.

3.2. Geometry and Boundary Conditions

A radius vector R = R (q_1, q_2) is used to describe the mid-surface of an arbitrary shell, R = R (q_1, q_2), where q_1, q_2 form an orthogonal coordinate system. For a toroidal shell with circular cross-section shown in Figure 2.2, the Lamé parameters of the shell are represented by A_1, A_2, and the radii of curvatures by R_1, R_2. These four parameters satisfy the following equations [Soedel(2004)]

$$A_1 = R + r\cos\theta; A_2 = r; R_1 = (R + r\cos\theta)/\cos\theta; R_2 = r \tag{3.1}$$

where $q_2 \equiv \theta$ is the meridional angle, measured clockwise from the positive horizontal, R is the bend radius, and r is the radius of the cross-section.

For a complete toroidal shell, the meridian forms a closed curve and the boundary conditions are not considered. For a toroidal shell with incomplete meridian, the current method can be extended by specifying appropriate boundary conditions at the ends of the meridian of such a shell.

3.3. Shear Deformation Theory

In the vibration analysis of a toroidal shell, the effect of the shear deformation can be neglected in the case for thin-walled shells. When the shell thickness increases, the shear deformation becomes an essential factor to determine their vibration characteristics. The shear deformation theory presented by [Soedel(1982), (2004)] is adopted in this section. The assumptions of this theory are in accordance with those of the Timoshenko beam theory and the Mindlin plate theory, in which the normal stress direction is zero in the thickness, and a normal to the shell mid-surface remain straight, but not necessarily normal. The theory discussed here is applied for the linear behavior for general moderately thick shells with elastic, homogeneous and isotropic materials.

3.3.1. Strain-Displacement Relationships

In the shear deformation theory, ε_1, ε_2 and ε_3 are the strain components, and κ_1, κ_2 and κ_3 are the bending strains on the mid-surface. The mid-surface strains and changes of curvature given in this theory are [Soedel(1982), (2004)]:

$$\begin{aligned}&\varepsilon_1 = f_1 u_{1,1} + f_2 u_2 + f_3 u_3; \varepsilon_2 = f_4 u_1 + f_5 u_{2,2} + f_6 u_3; \varepsilon_{12} = f_7 u_{1,2} + f_8 u_1 + f_9 u_{2,1} + f_{10} u_2,\\&\kappa_1 = f_{11}\beta_{1,1} + f_{12}\beta_2; \kappa_2 = f_{13}\beta_1 + f_{14}\beta_{2,2}; \kappa_{12} = f_{15}\beta_{1,2} + f_{16}\beta_1 + f_{17}\beta_{2,1} + f_{18}\beta_2,\end{aligned} \tag{3.2}$$

where u_1, u_2, u_3 are the mid-surface displacements in the q_1, q_2 and normal directions, respectively, β_1, β_2 are the mid-surface rotations about the local q_1, q_2 directions, respectively. The comma subscripts represent the differentiation with respect to the q_i variables which follows.

The transverse shear strains are given by

$$\varepsilon_{13} = f_{19} u_1 + f_{20} u_{3,1} + \beta_1; \varepsilon_{23} = f_{21} u_2 + f_{22} u_{3,2} + \beta_2, \tag{3.3}$$

where f_i are the coefficients of the strain components and changes of curvature defined in terms of the Lamé parameters and the radii of curvatures by [Soedel(1982), (2004)]:

$$\begin{aligned}&f_1 = 1/A_1; f_2 = A_{1,2}/(A_1 A_2); f_3 = 1/R_1; f_4 = A_{2,1}/(A_1 A_2); f_5 = 1/A_2; f_6 = 1/R_2;\\&f_7 = f_5; f_8 = -A_{1,2}/(A_1 A_2); f_9 = f_1; f_{10} = -f_4; f_{11} = f_1; f_{12} = f_2; f_{13} = f_4; f_{14} = f_5;\\&f_{15} = f_5; f_{16} = -f_2; f_{17} = f_1; f_{18} = -f_4; f_{19} = -f_3; f_{20} = f_1; f_{21} = -f_6; f_{22} = f_5.\end{aligned} \tag{3.4}$$

The force and moment resultants are given by

$$\begin{aligned}&N_1 = g_1\varepsilon_1 + g_2\varepsilon_2; N_2 = g_3\varepsilon_1 + g_4\varepsilon_2; N_{12} = g_5\varepsilon_{12},\\&M_1 = g_6\kappa_1 + g_7\kappa_2; M_2 = g_8\kappa_1 + g_9\kappa_2; M_{12} = g_{10}\kappa_{12},\end{aligned} \tag{3.5}$$

where the transverse shear resultants are

$$Q_{13} = g_{11} u_1 + g_{12} u_{3,1} + g_{13}\beta_1; Q_{23} = g_{14} u_2 + g_{15} u_{3,2} + g_{16}\beta_2, \tag{3.6}$$

The coefficients are defined in terms of the materials and geometric properties by

$$g_1 = K; g_2 = vK; g_3 = g_2; g_4 = g_1; g_5 = (1-v)K/2; g_6 = D; g_7 = vD; g_8 = g_7, \quad (3.7)$$
$$g_9 = g_6; g_{10} = (1-v)D/2; g_{11} = -k'Gh/R_1; g_{12} = -k'Gh/A_1; g_{13} = k'Gh,$$
$$g_{14} = -k'Gh/R_2; g_{15} = k'Gh/A_2; g_{16} = g_{13}; g_{17} = A_1A_2k'Gh; g_{18} = g_{13},$$

where E is the Young's Modulus, v the Poisson's ratio, k' the shear factor (taken as 2/3), h the shell wall thickness, $K=Eh/(1-v^2)$, $D=Kh^2/12$, and $2G=E/(1+v)$.

3.3.2. Equilibrium Equations

The equilibrium equations are based on [Soedel(1982), (2004)]

$$h_1N_{1,1} + h_2N_1 + h_3N_2 + h_4N_{12,2} + h_5N_{12} + h_6u_1 + h_7u_{3,1} + h_8\beta_1 + h_9u_{1,tt} = 0, \quad (3.8)$$
$$h_{10}N_1 + h_{11}N_{2,2} + h_{12}N_2 + h_{13}N_{12,1} + h_{14}N_{12} + h_{15}u_2 + h_{16}u_{3,2} + h_{17}\beta_2 + h_{18}u_{2,tt} = 0,$$
$$h_{19}N_1 + h_{20}N_2 + h_{21}u_{1,1} + h_{22}u_1 + h_{23}u_{2,2} + h_{24}u_2 + h_{25}u_{3,11} + h_{26}u_{3,1} + h_{27}u_{3,2} + h_{28}u_{3,2}$$
$$+ h_{29}\beta_{1,1} + h_{30}\beta_1 + h_{31}\beta_{2,2} + h_{32}\beta_2 + h_{33}u_{3,tt} = 0,$$
$$h_{34}M_{1,1} + h_{35}M_1 + h_{36}M_2 + h_{37}M_{12,2} + h_{38}M_{12} + h_{39}u_1 + h_{40}u_{3,1} + h_{41}\beta_1 + h_{42}\beta_{1,tt} = 0,$$
$$h_{43}M_1 + h_{44}M_{2,2} + h_{45}M_2 + h_{46}M_{12,1} + h_{47}M_{12} + h_{48}u_2 + h_{49}u_{3,2} + h_{50}\beta_2 + h_{51}\beta_{2,tt} = 0,$$

where t is time. It is noted that coefficients h_{42} and h_{51} associated with rotary inertia are included in Eq. (3.8d) and Eq. (3.8e). The other coefficients h_i defined in terms of the material and geometric properties are given by

$$h_1 = -A_2; h_2 = -A_{2,1}; h_3 = -h_2; h_4 = -A_1; h_5 = -2A_{1,2}; h_6 = g_{17}/R_1^2; h_7 = -g_{17}/(A_1R_1); \quad (3.9)$$
$$h_8 = -g_{17}/R_1; h_9 = A_1A_2\gamma h; h_{10} = -A_1A_2; h_{11} = A_{1.2}; h_{12} = h_4; h_{13} = -h_{11}; h_{14} = h_1;$$
$$h_{15} = -2A_{2,1}; h_{16} = g_{17}/R_2^2; h_{17} = -g_{17}/(A_2R_2); h_{18} = -g_{17}/R_2; h_{19} = h_9; h_{20} = h_{10};$$
$$h_{21} = A_1A_2/R_1; h_{22} = A_1A_2/R_2; h_{23} = -g_{18}A_2/R_1; h_{24} = g_{18}(A_2/R_1)_{,1}; h_{25} = g_{18}A_1/R_2;$$
$$h_{26} = g_{18}(A_1/R_2)_{,2}; h_{27} = -g_{18}A_2/A_1; h_{28} = -g_{18}(A_2/R_1)_{,1}; h_{29} = -g_{18}A_1/A_2;$$
$$h_{30} = -g_{18}(A_1/A_2)_{,2}; h_{31} = -g_{18}A_2; h_{32} = -g_{18}A_{2,1}; h_{33} = -g_{18}A_1; h_{34} = -g_{18}A_{1,2};$$
$$h_{35} = h_9; h_{36} = h_{10}; h_{37} = -h_1; h_{38} = -h_2; h_{39} = -h_2; h_{40} = -h_{12}; h_{41} = 2h_{11}; h_{42} = -h_8;$$
$$h_{43} = -h_{17}/A_1; h_{44} = -h_{17}; h_{45} = -A_1A_2\gamma h^3/12; h_{46} = h_{13}; h_{47} = -h_4; h_{48} = h_{11};$$
$$h_{49} = -h_1; h_{50} = 2g_3; h_{51} = g_{17}/R_2; h_{52} = -g_{17}/A_2; h_{53} = -g_{17}; h_{54} = h_{45};$$

where γ is the mass density. The theory will be further written for shells of revolution. The variables q_1 and q_2 are respectively taken to represent the circumferential direction φ, and the meridional direction θ. Adopting a modal approach, the displacement components and rotations can be expanded in a Fourier series in the circumferential direction as

$$u_1 = u(\theta)\sin m\varphi \sin \omega t; u_2 = v(\theta)\cos m\varphi \sin \omega t; u_3 = w(\theta)\cos m\varphi \sin \omega t, \quad (3.10)$$
$$\beta_1 = \alpha(\theta)\sin m\varphi \sin \omega t; \beta_2 = \beta(\theta)\cos m\varphi \sin \omega t,$$

where m is the circumferential harmonic number, ω is the circular frequency in rad/sec, u, v, and w are the amplitudes of the displacements in the circumferential, meridional and normal directions for the mth harmonic, α and β are the amplitudes of the respective rotations. For each choice of m, the problem can be reduced mathematically to one-dimension.

The equations of motion for shells of revolution can be represented as

$$\begin{aligned}
&l_1u_{,22}+l_2u_{,2}+(l_3-m^2l_4)u-ml_5v_{,2}-ml_6v-ml_7u_3+l_8\alpha-l_9\omega^2u=0,\\
&-ml_{10}u_{,2}-ml_{11}u+l_{12}v_{,22}+l_{13}v_{,2}+(l_{14}-m^2l_{19})v+l_{16}w_{,2}+l_{17}w+l_{18}\beta-l_{19}\omega^2v=0,\\
&ml_{20}u+l_{21}v_{,2}+l_{22}v+l_{23}w_{,22}+l_{24}w_{,2}+(l_{25}-m^2l_{26})w+ml_{27}\alpha+l_{28}\beta_{,2}\\
&+l_{29}\beta-l_{30}\omega^2w=0,\\
&l_{31}u-ml_{32}w+l_{33}\alpha_{,22}+l_{34}\alpha_{,2}+(l_{35}-m^2l_{36})\alpha-ml_{37}\beta_{,2}-ml_{38}\beta-l_{39}\omega^2\alpha=0,\\
&l_{40}v+l_{41}w_{,2}+ml_{42}\alpha_{,2}+ml_{43}\alpha+l_{44}\beta_{,22}+l_{45}\beta_{,2}+(l_{46}-m^2l_{47})\beta-l_{48}\omega^2\beta=0,
\end{aligned} \tag{3.11}$$

where the l_i are known coefficients expressed in terms of the geometric parameters and the coefficients f_i, g_i and h_i given in this section. It is noted that for each circumferential mode m, there are five field equations, in one geometric variable, for five unknown functions and unknown frequency ω. The m=0 case, i.e. the axi-symmetric circumferential harmonic described in the preceding, the displacement component u_1, the rotation β_1, and the resultants N_{12}, M_{12}, Q_1 will be zero, and the Eq. (3.11a) and Eq. (3.11d) will become trivial.

3.4. Differential Quadrature Method

As discussed previously, the advantage of using DQM is to convert the set of differential equations with a particular harmonic m to a set of simultaneous equations [Shu(2000), Bert and Malik(1996)]. In the case of a shear deformation problem, a one-dimensional grid of sampling points is defined along a meridian of the radial plane. Then the derivative of a function in a given direction is replaced by the weighted sum of the values of the function at specified sampling points in a line following the given direction.

For a generic function $f(x)$ of a single variable, the series used to replace the rth derivative of the function at the sampling point x_i is taken as

$$\frac{d^{(r)}f(x)}{dx^r}\Big|_{x_i}=\sum A^{(r)}{}_{ih}f(x_h) \tag{3.12}$$

where the $A^{(r)}{}_{ih}$ weighting coefficients of the rth order derivative in the x direction for the ith sampling point, $f(x_h)$ is the value of f(x) at the sampling point position x_h. The number of sampling points in the x direction is denoted by M.

In the case of a generic function with two variables $g(x,y)$, the series for the $(r+s)$th partial derivative at the sampling point x_i, y_j is taken as

$$\frac{\partial^{(r+s)}g(x,y)}{\partial x^r\partial y^s}\Big|_{x_i,y_j}=\sum A^{(r)}{}_{ih}\sum B^{(s)}{}_{jk}g(x_h,y_k) \tag{3.13}$$

where $g(x_h, y_k)$ is the value of $g(x,y)$ at the sampling point position x_h, y_k. $B^{(s)}{}_{jk}$ describes the series for the y direction, and the number of sampling points in the x direction is denoted by N. In the shear deformation analysis of the present geometry with a complete meridian, a Fourier harmonic basis was selected for the weighting coefficients [Bert and Malik(1996)], and sampling points were equally spaced along the coordinate q_2.

To apply the quadrature rule of Eq. (3.12) of the DQM to the differential equation of the shear deformation theory results to a set of linear simultaneous equations with the following form

$$\begin{aligned}
&L_{11}u + L_{12}v + L_{13}w + L_{14}\alpha + L_{15}\beta + \gamma\omega^2 u = 0,\\
&L_{21}u + L_{22}v + L_{23}w + L_{24}\alpha + L_{25}\beta + \gamma\omega^2 v = 0,\\
&L_{31}u + L_{32}v + L_{33}w + L_{34}\alpha + L_{35}\beta + \gamma\omega^2 w = 0,\\
&L_{41}u + L_{42}v + L_{43}w + L_{44}\alpha + L_{45}\beta + \gamma\omega^2 \alpha = 0,\\
&L_{51}u + L_{52}v + L_{53}w + L_{54}\alpha + L_{55}\beta + \gamma\omega^2 \beta = 0.
\end{aligned} \tag{3.14}$$

3.5. Validation and Results

The accuracy of current results for the natural frequencies based on SDT using DQM are verified by comparison with results based on thick shell Lagrangian (TSL) from Kosawada et al (1986) and the FEM results from Buchanan and Liu (2005). Six case studies with their geometric properties are presented in Table 3-1, where r_i, r_m, and r_o represent respectively the inside, mean, and outside radius of the cross-section. Cases 1-3 are shells having radius to thickness ratios of r_m/h about 6.5, and are considered moderately-thick shells. Cases 4-6 with r_m/h ratio less than 4 are represented thick shells. Two different material properties (Mat.1 and Mat.2) are used to verify the results and the comparison results are provided in Tables 3-2 to 3-3.

Mat.1: $E = 0.207 \times 10^{12}$ Pa; $v = 0.3$; $\gamma = 7800\,\text{kg/m}^3$,

Mat.2: $E = 0.193 \times 10^{12}$ Pa; $v = 0.291$; $\gamma = 7850\,\text{kg/m}^3$,

Table 3.1. Geometry properties of six toroidal shells

Case	r_i	r_o	h	r_m	R	r_m/h	R/r_m	material
1	0.9225	1.0775	0.1549	1	20	6.46	20	1
2	0.9225	1.0775	0.1549	1	10	6.46	10	1
3	0.9225	1.0775	0.1549	1	6.67	6.46	6.67	1
4	0.75	1	0.25	0.875	2.5	3.50	2.86	2
5	0.50	1	0.50	0.750	2.5	1.50	3.33	2
6	0.25	1	0.75	0.625	2.5	0.83	4.00	2

Table 3.2. Comparison of SDT frequencies with ω_i (Hz) TSL results of Kosawada et al(1986) for moderately thick shell – validation cases 1-3 ($m = 2$)

Mode	Case 1		Case 2		Case 3	
	TSL	SDT	TSL	SDT	TSL	SDT
1	3.6	3.9	13.2	13.5	25.1	26.6
2	-	4.0	-	13.6	-	26.9
3	57.6	58.0	105.2	104.2	120.8	122.0
4	-	91.3	-	116.0	-	129.1
5	105.8	105.4	128.0	128.1	180.2	182.1
6	-	105.7	-	180.1	-	263.0

Table 3.3. Comparison of non-dimensional SDT frequencies Ω_i with FEM results of Buchanan and Liu (2005) (Table 6, p. 258) for thick shells – validation cases 4-6 ($m = 2$)

Mode	Case 4		Case 5		Case 6	
	FEM	SDT	FEM	SDT	FEM	SDT
1	0.2482	0.2491	0.2568	0.2646	0.2537	0.2814
2	0.2857	0.2832	0.2965	0.2871	0.2975	0.2818
3	0.5288	0.5243	0.9762	0.9506	1.0081	1.0072
4	0.5396	0.5361	1.0553	0.9676	1.2172	1.2529
5	0.9655	0.9633	1.0782	1.0058	1.7606	1.4237
6	1.1137	1.1146	1.1851	1.1946	1.7729	1.4357

In Table 3-2, the natural frequencies with mode numbers based on SDT using DQM (current results) are validated by comparing results based on thick shell Lagrangian (TSL) from Kosawada et al (1986), where only the symmetric modes were represented in their study. It is observed that there is generally good agreement within 2% between the frequencies for these three cases. It is also seen that the largest difference can be evaluated in the shell with the smallest frequency and the relative error due to scaling is the largest.

In Table 3-3, the current SDT results are verified by the FEM results given by Buchanan and Liu (2005). It is found that there is agreement within 1%, 3%, and 11% for the fundamental frequency of cases 4-6. The largest difference occurs for the thickest shell (case 6). It is seen that although the R/r_o is constant for these cases, the R/r_m ratio is not. In general, the results of Tables 3-2 to 3-3 indicate that SDT shows close agreement with other applicable theory for an r_m/h ratio ranging from about 100 to 3.

4. Elasticity Theory

Hollow tori of revolution (thick toroidal shells) have found a variety of engineering applications in submarine hulls, pressure vessels, gyroscopes, piping, and machinery, ect. These toroidal shell structures can carry external and internal dynamic pressure loadings. This is especially the case when the wall of the shell is considered thick, and the assumption that a normal is no longer valid. Three-dimensional theory of elasticity is required to analyze for such bodies and applications. This section will focus on free vibration analysis for thick toroidal shells using elasticity theory.

4.1. Background

There are a number of studies focused on three-dimensional vibration analyses of thick shells of revolution. Among the geometries that have been considered are thick-walled cylinders [Armenakas et al(1969), Chou and Achenback(1972), Singal and Williams(1988), So and Leissa(1997), Liew et al(2000), Zhou et al(2003), McGee and Kim(2010)], thick-walled spherical [Gautham and Ganesan(1994), Ding and Chen(1996), Chen(1999), Chen and Ding(2001), Buchanan and Rich(2002), Leissa and Kang(1999), (2000)] [Kang and Leissa(2000), (2006)], thick-walled conical [Leissa and So(1995), Leissa and Kang(1999)], thick-walled paraboloidal [Kang and Leissa(2005)]. Vibration analysis for thick toroidal shells to date have been developed using FEM [Buchanan and Liu(2005)], or have been studied only with axi-symmetric vibration [McGill and Lenzen(1967), Jiang and Redekop(2002)]. A distinguishing feature of the present section is to establish effective theory to develop equations that predict the natural frequencies of vibration of thick shells of revolution that is specifically applicable to cylindrical, spherical and toroidal shells. The restriction considered in the study of this section is for the linear behaviour of shells with elastic homogeneous isotropic materials.

4.2. Geometry and Boundary Conditions

The toroidal shell has a bend radius of R, and is bounded by inner and outer circumferential surfaces of cross-sectional radii $r=r_i$ and $r=r_o$ respectively (Figure 4.1).

The shell is complete in the meridional and circumferential directions, θ and φ. By assuming the mode shape in the circumferential direction, the mathematical solution is reduced to the two dimensions of the cross-section, r and θ. In the case of the theory of elasticity, the boundary conditions must be satisfied on both the inner and outer surface to ensure zero traction conditions.

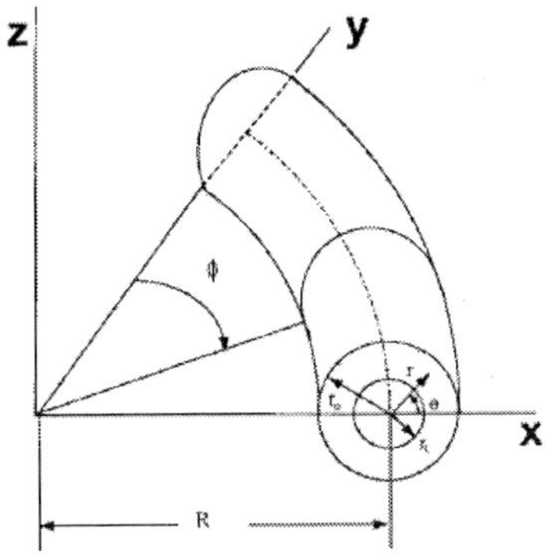

Figure 4.1. View showing first quadrant of a thick-walled toroidal shell.

4.3. Elasticity Theory

In the case of thick toroidal shells, the three-dimensional theory of elasticity must be applied to analyze their static or dynamic displacements and stresses, and free vibrations. Compared with the equations of motion based upon various thin shell theories, the equations of motion based on three-dimensional theory of elasticity are more complex and more thorough to provide accuracy.

4.3.1. Strain-displacement relationships

In the elasticity theory, the normal and shear strain components can be expressed as [Redekop(1992), Buchanan and Liu(2005)]

$$\begin{aligned}
&\varepsilon_1 \equiv \varepsilon_\phi = f_1 u_{1,1} + f_2 u_2 + f_3 u_3; \varepsilon_4 \equiv \varepsilon_{\theta\phi} = f_6 u_{1,2} + f_7 u_1 + f_8 u_{2,1}, \\
&\varepsilon_2 \equiv \varepsilon_\theta = f_4 u_{2,2} + f_5 u_3; \varepsilon_5 \equiv \varepsilon_{r\phi} = u_{1,3} + f_9 u_1 + f_{10} u_{3,1}, \\
&\varepsilon_3 \equiv \varepsilon_r = u_{3,3}; \varepsilon_6 \equiv \varepsilon_{r\theta} = u_{2,3} + f_{11} u_2 + f_{12} u_{3,2},
\end{aligned} \tag{4.1}$$

where u_1, u_2, u_3 are respectively the displacement components in the $q_1 \equiv \varphi$, $q_2 \equiv \theta$, and $q_3 \equiv r$ directions (Figure 4.1), and the comma subscript indicates differentiation with respect to the q_i variable that follows.

The coefficients f_i in the expressions for the strain components are defined for toroidal coordinates as

$$\begin{aligned}
&f_1 = 1/\rho; f_2 = -\sin\theta/\rho; f_3 = \cos\theta/\rho; f_4 = 1/r; f_5 = f_4; f_6 = f_4; \\
&f_7 = -f_2; f_8 = f_1; f_9 = -f_3; f_{10} = f_1; f_{11} = -f_4; f_{12} = f_4;
\end{aligned} \tag{4.2}$$

where $\rho = R + r\cos\theta$, R is again the bend radius, and r is now the radial coordinate.

4.3.2. Equilibrium Equations

The normal and shear stress components are given by [Redekop(1992), Buchanan and Liu(2005)]

$$\begin{aligned}
&\sigma_1 \equiv \sigma_\phi = g_1\varepsilon_1 + g_2\varepsilon_2 + g_3\varepsilon_3; \sigma_4 \equiv \sigma_{\theta\phi} = g_{10}\varepsilon_4; \\
&\sigma_2 \equiv \sigma_\theta = g_4\varepsilon_1 + g_5\varepsilon_2 + g_6\varepsilon_3; \sigma_5 \equiv \sigma_{r\phi} = g_{11}\varepsilon_5; \\
&\sigma_3 \equiv \sigma_r = g_7\varepsilon_1 + g_8\varepsilon_2 + g_9\varepsilon_3, \sigma_6 \equiv \sigma_{r\theta} = g_{12}\varepsilon_6,
\end{aligned} \tag{4.3}$$

The coefficients g_i are defined for homogeneous isotropic materials by

$$g_1 = g_5 = g_9 = K_1; g_2 = g_3 = g_4 = g_6 = g_7 = g_8 = K_2; g_{10} = g_{11} = g_{12} = K_3, \tag{4.4}$$

where $K_1 = 2(1-v)G/(1-2v)$, $K_2 = 2vG/(1-2v)$, $K_3 = 2G$, and E is the Young's modulus, v the Poisson's ratio, and $2G = E/(1+v)$.

The equations of motion for a toroidal shell are given by

$$h_1\sigma_{1,1}+h_2\sigma_{4,2}+h_3\sigma_4+\sigma_{5,3}+h_4\sigma_5=\gamma u_{1,tt},$$
$$h_5\sigma_1+h_6\sigma_{2,2}+h_7\sigma_2+h_8\sigma_{4,1}+\sigma_{6,3}+h_9\sigma_6=\gamma u_{2,tt},$$
$$h_{10}\sigma_1+h_{11}\sigma_2+\sigma_{3,3}+h_{12}\sigma_3+h_{13}\sigma_{5,1}+h_{14}\sigma_{6,2}+h_{15}\sigma_6=\gamma u_{3,tt}, \tag{4.5}$$

where γ is the mass density. The coefficients h_i are defined for toroidal coordinates as

$$h_1=1/\rho; c_2=1/r; h_3=-2\sin\theta/\rho; h_4=(R+3r\cos\theta)/r\rho; h_5=\sin\theta/\rho,$$
$$h_6=h_2; h_7=-h_5; h_8=h_1; h_9=(2R+3r\cos\theta)/r\rho; h_{10}=-\cos\theta/\rho,$$
$$h_{11}=-h_2; h_{12}=(R+2r\cos\theta)/r\rho; h_{13}=-h_1; h_{14}=h_2; h_{15}=-h_5, \tag{4.6}$$

A modal approach is used in this analysis and then the displacement components can be expanded in a Fourier series in the circumferential directions as

$$u_1=u(\theta,r)\sin m\varphi\sin\omega t; u_2=v(\theta,r)\cos m\varphi\sin\omega t; u_3=w(\theta,r)\cos m\varphi\sin\omega t, \tag{4.7}$$

where m is the number of the circumferential harmonic, u, v, and w are the amplitudes of the displacements in the circumferential, meridional and normal directions for the mth harmonic.

Now functions contain both θ and r, and ω is the circular frequency in rad/sec and each harmonic may be analyzed separately, and the problem can be mathematically reduced to two-dimensions.

The equations of motion for a solid of revolution can be represented as [Lur'e(1964)]

$$l_1u_{,22}+l_2u_{,2}+l_3u_{,33}+l_4u_{,3}+(l_5-m^2l_6)u-ml_7v_{,2}-ml_8v-ml_9w_{,3}-ml_{10}w-\gamma u_{,tt}=0,$$
$$ml_{11}u_{,2}+ml_{12}u+l_{13}v_{,22}+l_{14}v_{,2}+l_{15}v_{,33}+l_{16}v_{,3}+(l_{17}-m^2l_{18})v+l_{19}w_{,2}+l_{20}w_{,3}$$
$$+l_{21}w_{,23}+l_{22}w-\gamma v_{,tt}=0,$$
$$ml_{23}u_{,3}+ml_{24}u+l_{25}v_{,2}+l_{26}v_{,3}+l_{27}v_{,23}+l_{28}v+l_{29}w_{,22}+l_{30}w_{,2}+l_{31}w_{,33}+l_{33}w_{,3}$$
$$+(l_{33}-m^2l_{34})w-\gamma w_{,tt}=0, \tag{4.8}$$

where the l_i are known coefficients determined in terms of the geometric parameters and the coefficients f_i, g_i, and h_i given in this section. For each circumferential mode m, there are three field equations, in two geometric variables, for three unknown functions and the unknown frequency ω. The Eq. (4.8) can be applied to the interior points of the domain, whereas on the inner and outer surfaces, the following boundary conditions apply: $\sigma_{31}=0$, $\sigma_{32}=0$, and $\sigma_{33}=0$.

In the case of the axi-symmetric circumferential harmonic, i.e. $m=0$, the displacement component u_1, and the stresses σ_1, σ_4, and σ_5 will be zero, and the Eq. (4.8a) will become trivial.

4.4. Differential Quadrature Method

In the present study, the problem is solved again using the DQM [Shu(2000)]. In the application of the method to the three-dimensional elasticity problem, a two-dimensional grid is defined on the cross-sectional area of the radial plane. Following the same quadrature rule of Eq. (3.12) of the DQM to the differential equation of present problem, the equations reduce to the following form

$$L_{11}u + L_{12}v + L_{13}w + \gamma\omega^2 u = 0,$$
$$L_{21}u + L_{22}v + L_{23}w + \gamma\omega^2 v = 0, \quad (4.9)$$
$$L_{31}u + L_{32}v + L_{33}w + \gamma\omega^2 w = 0.$$

4.5. Validation and Results

The accuracy of the natural frequencies based on ELT using DQM are validated by comparing results based on thick shell FEM results from Buchanan and Liu (2005). Three thick shell cases with their geometric properties are given in Table 4-1, where r_i, r_m, and r_o represent the inside, mean, and outside radius of the cross-section, respectively. The material properties used in this study is given as follows. The comparison results are given in Table 4-2.

$$E = 0.193 \times 10^{12}\ \text{Pa};\ \nu = 0.291;\ \gamma = 7850\ \text{kg/m}^3.$$

Table 4.1. Geometry properties of three toroidal shell validation cases

Case	r_i	r_o	h	r_m	R	r_m/h	R/r_m
1	0.75	1	0.25	0.875	2.5	3.50	2.86
2	0.50	1	0.50	0.750	2.5	1.50	3.33
3	0.25	1	0.75	0.625	2.5	0.83	4.00

Table 4.2. Comparison of non-dimensional ELT frequencies Ω_i with FEM results of Buchanan and Liu (2005) (Table 6, p. 258) for thick shells ($m = 2$)

Mode	Case 1		Case 2		Case 3	
	FEM	ELT	FEM	ELT	FEM	ELT
1	0.2482	0.2479	0.2568	0.2567	0.2537	0.2534
2	0.2857	0.2854	0.2965	0.2963	0.2975	0.2973
3	0.5288	0.5263	0.9762	0.9759	1.0081	1.0080
4	0.5396	0.5371	1.0553	1.0529	1.2172	1.2172
5	0.9655	0.9654	1.0782	1.0758	1.7606	1.7535
6	1.1137	1.1134	1.1851	1.1847	1.7729	1.7655

In Table 4-2, it is observed that there is a close agreement of FEM and ELT results are within 1% for all three cases and modes. It is also seen that the trend of the fundamental frequency from thinnest to thickest shell (case 1 to case 3) is a very modestly increasing one.

5. Finite Element Method

The FEM is a very powerful technique that can be used to predict the natural frequencies together with their corresponding mode shapes. In this section, the commercial FEM program ADINA [ADINA(2003), Bathe(1996)] is used as an alternate solution to predict the vibration characteristic for thin and thick curved tubes and toroidal shells with emphasis on two standard geometries of interest to engineers.

5.1. Background

In recent years, there has been substantial progress in the study of the vibration of curved tubes and toroidal shells [Redekop(1994a-c), (1995), (1997), Huang et al(1997), Ornyak et al(2007)]. Continuing contributions and new applications of these structures can be seen in the engineering practice of fusion reactor vessels, nuclear piping systems, and satellite antennas, ect. There is extensive research available on vibration analysis for both thin-walled (discussed in section 2) and thick-walled geometries (discussed in sections 3 and 4). The analysis of the present section is conducted using the FEM to obtain the results given for thin and thick curved tubes extending through a circumferential angle α of 90° or 180°, and for toroidal shells ($\alpha = 360^{o}$).

For the FEM program ADINA [ADINA(2003), Bathe(1996)], the analysis for thin shells is based on the Love-Kirchoff while the analysis for the thick shells is based on the theory of elasticity. In this section, new results obtained using this FEM are for two standard components, a '5s' thin shell and a '80s' thick one. Each of these standard components forms a 90° curved tube (elbow), two welded together compose an 180° curved tube, and four welded together assemble a complete toroidal shell. The vibration results presented herein for these geometries can be used as benchmarks for researchers and as reference values for design engineers.

5.2. FEM

As discussed at the beginning of this section, the FEM program ADINA [ADINA(2003), Bathe(1996)] is used as the main tool to determine the natural frequencies for curved tubes and toroidal shells. A 4-node 24-degree-of-freedom shell element is used for the analysis of thin curved tubes and shells. This element is based on the shear deformation shell theory in the ADINA program. A 27-node 81-degree-of-freedom brick element is utilized to analyze the thick curved tubes and shells and this element is based on the theory of elasticity.

The modeling of FEM calculation accounts for the full geometry for each case study, with no account for geometric symmetry. The FEM results are calculated based on the finest meshes that could be run on a computer system. Boundary conditions considered are of the fixed, diaphragm, and completely free types.

For the free vibration analysis [Bathe(1996)], the numerical solution is reduced to solving the problem

$$[K]\,(u) = \lambda\,[M]\,(u) \tag{7.1}$$

where [K] and [M] are the stiffness and mass matrices, respectively, and (u) and λ are the modal vector and the frequency parameter.

Generally, the first six natural frequencies are of practical interest and they are determined in the current analysis. In the ADINA program, the eigenvalue analysis is carried out using the subspace iteration method.

5.3. Validation and New Results for Thin Shells

The verification and the new results obtained using the FEM analysis for thin curved tubes and toroidal shells are discussed in the following subsections.

5.3.1. Validation Using the DQM

A DQM solution and its accuracy presented for the vibration analysis of thin toroidal shells have been discussed in section 2. This solution is based on Sanders-Budiansky thin shell theory which is of the Love-Kirchoff type. It has been adapted as well for the vibration analysis of thin curved tubes of circumferential angle less than 360° [Redekop and Muhammad(2003)].

The accuracy of the current FEM method for thin curved tubes is validated with the DQM solution as well as the previously published results [Ming(2002)]. The material and geometry properties are given in Table 5-1 [Wang et al(2006)].

Table 5.1. Data for the material and geometry of the curved tube

Parameter	Values
Elastic modulus: E (Pa)	0.207×10^{12}
Poisson's ratio: v	0.3
Density: ρ (kg/ m^3)	7800
Thickness : h (m)	0.003
Radius of the cross-section: r (m)	0.1
Bend radius: R (m)	1.0
r/h ratio: r/h	33.3
R/r ratio: R/r	10.0
Pipe factor: $\xi = r^2/(hR)$	3.33

The six lowest frequencies for the thin curved tubes with a diaphragm and fixed end supports are given in Table 5-2 including the previous published results [Ming(2002)], current FEM results and DQM results. There is generally good agreement among these three results.

It is found that the DQM agrees to FEM within 0.7%. The close agreement indicates that FEM can be used as an alternate solution to predict the vibration characteristic and it can enhance the efficiency of the analysis process.

Table 5.2. Validation study of frequencies (Hz) for thin 90° curved pipes

Support	Diaphragm ends			Fixed ends		
Mode	Ming (2002)	DQM	FEM	Ming (2002)	DQM	FEM
1	67.96	66.98	67.41	347.76	334.64	336.73
2	71.50	70.98	71.43	530.21	506.64	509.54
3	277.09	275.04	276.33	539.10	507.77	510.72
4	277.15	275.05	276.34	540.08	527.39	530.60
5	470.67	467.24	468.51	548.21	533.52	536.71
6	471.00	467.56	468.84	677.46	669.26	673.02

5.3.2. New Results Using FEM

The new results for natural frequencies of thin curved tubes and a toroidal shell using FEM, ADINA program are presented and the material and geometry properties are given in Table 5-3 [Wang et al(2006)].

These parameters associate with a commercial product with the '5s' designation. The results carried out are for 90° and 180° curved tubes with fixed and free end supports, and for a completely free toroidal shell. The first six mode shapes for these five tube and shell cases, as determined by using FEM, are presented in Table 5-4 [Wang et al(2006)].

Table 5.3. Data for the material and geometry of the thin curved pipe

Parameter	Values
Elastic modulus: E (Pa)	0.193×10^{12}
Poisson's ratio: ν	0.291
Density: ρ (kg/ m^3)	7850
Thickness : h (m)	0.00211
Radius of the cross-section: r (m)	0.056095
Bend radius: R (m)	0.1524
r/h ratio: r/h	26.59
R/r ratio: R/r	2.717
Pipe factor: $\xi = r^2/(hR)$	9.785

Table 5.4. Frequencies (Hz) for '5s' curved pipes and toroidal shell

Support	Fixed ends		Completely free		
α	90°	180°	90°	180°	360°
mode	FEM	FEM	FEM	FEM	FEM
1	3152	1279	526	464	659
2	3422	2220	533	591	845
3	3482	2386	623	612	845
4	3614	2437	628	613	935
5	3962	2450	1359	724	935
6	4103	2466	1363	724	1639

5.4. Validation and New Results for Thick Shells

The literature study on the vibration analysis for thick toroidal geometries is relatively sparse. There is no known work on thick curved tubes, but a comprehensive study on thick toroidal shells, using FEM, has recently been published by Buchanan and Liu (2005) which we have demonstrated as a validation study in section 4 (Table 4-2). The new results obtained using FEM for both curved tubes and toroidal shells are demonstrated in the following.

5.4.1. New Results Using FEM

New results are given in Hz for the frequencies of thick curved tubes and a toroidal shell in this subsection. The material and geometry properties corresponding to a commercial product with the '80s' designation are given in Table 5-5 [Wang et al(2006)].

Table 5.5. Data for the material and geometry of the thick curved tube

Parameter	Values
Elastic modulus: E (Pa)	0.193×10^{12}
Poisson's ratio: v	0.291
Density: ρ (kg/ m^3)	7850
Internal cross-section radius: r_i (m)	0.04859
External cross-section radius: r_o (m)	0.05715
Bend radius: R (m)	0.1524
r_m/h ratio: r_m/h, r_m (mean cross-section radius)	6.176
R/r_m ratio: R/r_m, r_m (mean cross-section radius)	2.883
Pipe factor: $\xi = r^2/(hR)$	2.143

Table 5.6. Frequencies (Hz) for '80s' curved pipes and toroidal shell

Support	Fixed ends		Completely free			
α	90°	180°	90°	180°	360°	
mode	FEM	FEM	FEM	FEM	FEM	DQM
1	3547	1376	2038	1098	1620	1611
2	4385	2418	2051	1619	1693	1689
3	4947	2973	2171	1824	1693	-
4	5928	3179	2209	1995	1929	1924
5	6164	3553	3794	2272	1929	-
6	6486	3754	3804	2339	3048	3040

For the case of the commercial product with the '80s' designation in Table 5-5, the FEM results including curved 90° and 180° tubes with fixed and free end supports, and for a completely free toroidal shell are presented in Table 5-6. Also presented in Table 5-6 are the DQM results only available for the completely free shell. In the case of the toroidal shell, there is good agreement found within 0.6% between the results of FEM and DQM.

6. Pre-Stressed and Control Vibrations

All cases discussed so far are based on the free vibration analysis of toroidal shells without initial pressure. Over the years, significant progress has been made in the analysis of toroidal shells proposed for space telescopes, inflatable space structures (gossamer structures), neutron accelerators, space colonies, cooling tubes, etc. The dynamic behavior and the load carrying ability are the fundamental safety criterions in the design of such structures. In the study for free vibration of toroidal shells, it is imperative to obtain the natural frequencies and the mode shapes for such shells subjected to initial pressures. The summarized work in prestressed vibration analysis of toroidal shells, especially for gossamer structures is discussed in this section.

On the other perspective of such applications for toroidal shells, the effective vibration control always represent challenges both in theory and practice, and it has drawn a considerable attention in vibration and control research areas. Another goal of this section is to briefly summarize the control analyses of toroidal shells.

6.1. Pre-Stressed Vibration

The research for the free vibration of shell structures under the initial pressure can be traced back to the year of 1936, when Federhofer first studied the axially compressed circular cylindrical shell [Soedel(2004)]. The early work for toroidal shells subjected to initial pressure was Liepins (1965), who studied the free vibrations of prestressed toroidal shells. His research was based on the linearized shell theory and applied Fourier series expansions in the circumferential direction and finite difference approximations in the meridional direction. Jordan (1966) investigated the vibration and buckling of pressurized torus shells using Rayleigh quotient and also demonstrated the experiments' results. Later, Saigal et al (1986) presented the free vibration of a tire modeled as a toroidal membrane under internal pressure. In their study, they assumed the radius for the toroid which is much larger than the radius of the meridian circle and provided a simple model for obtaining a good first approximation of the free vibration response from a low pressure tire.

Since the 1990s, extremely lightweight and large inflatable toroidal structures, known as gossamer structures, have made strong pace of progress for space applications. Their extremely low weight, on-orbit deployability, and minimal stowage volume for launching are highly desirable characteristics for space applications, such as communication antennas, solar thermal propulsion, solar power in space, sun shades, solar sails, and many others. The vibration analysis with initial stress is essentially important for gossamer structures because they are subjected to a variety of dynamic loadings and there is extensive research work in this area.

On the analytical aspect, Lewis studied the free vibration of a prestressed toroidal shell using finite element analysis [Lewis(2000)]. His study demonstrated that the natural frequencies and mode shapes were significantly impacted by the aspect ratio of a toroidal shell and also the natural frequencies increased with increasing the internal pressure.

Jha et al (2002) investigated the free vibration analysis of an inflated torus with free boundary conditions and subjected to the initial pressure. The Sanders-Budiansky shell theory

[Sanders (1959), Budiansky (1968)] was used in their study with inclusions of geometric nonlinearity in the in-plane strains in conjunction with initial pressure. Jha and Inman (2003) presented vibration control of a gossamer toroidal structure using smart material actuators and sensors. In their study, based on Sanders' shell theory, they first developed the governing equations of motion for a toroidal shell subjected to the initial pressure using Galerkin's method. Then they continued carrying out the effects of the presence of piezoelectric patches on the toroidal shells. Jha and Inman (2004) studied the dynamic analysis of gossamer structures using Sanders' thin theory. They first developed the non-linear strain-displacement relation with the prestresses from three-dimensional elasticity theory. Then they derived the governing equations without any pressure and then modified them to include the effects of initial pressure. Their study demonstrated more accurate results on the vibration analysis of gossamer structures subjected to the internal pressure. On the aspect of the experimental analysis and FEM simulation, Griffith and Main (2000) demonstrated an experimental test for prestressed toroidal shells. In their test, a modified impact hammer is used to stimulate the global modes of the structure to avoid the local excitation. Park et al(2003) demonstrated the experimental testing to find resonant frequencies and mode shapes for gossamer toriodal structures and the testing results were validated using the finite element software ANSYS.

6.2. Control of Vibration

Recent development and analysis of smart materials and structures have demonstrated a number of new distributed actuators that can be spatially distributed over shell structures [Gabbert and Tzou(2001), Tzou and Bergman(1998)]. Research and experiments show that piezoelectric materials can be used as actuator or sensors to control structural configuration and to suppress some undesired vibrations. Recent work on distributed control characteristics and effectiveness has dealt mainly with plates, rings, cylindrical shells, and paraboloidal shells [Liu and Tzou(1998)] [Tzou et al(1994), (1996), (1997), (2002), (2003)] [Zhou and Tzo(2000)]. There have been a few studies on the control of vibrations for toroidal shells. Lewis and Inman (2001) presented the structural dynamics and vibration suppression via piezoelectric actuators of an inflatable torus using the computer software, ANSYS. In their study, the Eulaer-Bermoulli bean theory is used to model the interaction with the torus of the surface-mounted piezoelectric patches. Park et al(2001) studied the integration of smart materials into dynamics and control of inflatable space structures. In their study, an experimental investigation of vibration testing and control of an inflated thin-film torus was presented. Tzou and Wang (2002) presented the micro-sensing characteristics and modal voltages of linear and nonlinear toroidal shells. From their study, they demonstrated that the meridional membrane strains are the dominating signal component. In another study, Tzou and Wang (2003) presented dynamic and control analyses of toroidal shells. The fundamental constitutive relations of piezoelectricity were developed and control effectiveness of the parallel and the diagonal layouts were evaluated in their study. Jha and Inman (2002a) demonstrated the models of piezoelectric actuator and sensor for gossamer structures. In another study, they presented the vibration control for the gossamer toroidal structures using smart actuators and sensors [Jha and Inman(2002b)]. Later, they further investigated the optimal sizes and placements of piezoelectric actuators and sensors for gossamer toroidal structures [Jha and Inman(2003)]. They indicated the importance of optimizing the locations and sizes of the actuators to minimize the control effort in their study.

CONCLUSION

The research works on the vibration analysis of toroidal shells including the applications of different shell wall thicknesses associated with the appropriate theoretical approaches are summarized in this chapter. The vibration analysis of curved tubes and toroidal shells using FEM is also discussed. The research works to conduct the effects of initial pressure for the toroidal shell's vibration characteristic and the control properties for toroidal shells are reviewed in this chapter as well. The following are a set of conclusions formulated from this chapter:

1) Based on Sander-Budiansky shell theory, general equations are obtained to predict the vibration characteristics of thin walled isotropic shells of revolution with an arbitrary meridian.
2) Complete information about the study of natural frequencies for isotopic circular toroidal shells is determined and the excellent agreement among the DQM, FEM and RKM results are demonstrated.
3) General equations intended to predict the natural frequencies of a moderately-thick toroidal shell are developed based on the shear deformation theory of Soedel. The study indicates the value of the shear deformation theory for the vibration analysis of toroidal shells.
4) Based on the theory of elasticity, newer equations are obtained to evaluate the natural frequencies of thick toroidal shells.
5) The procedure using finite element method to demonstrate the vibration analysis for curved tubes and toroidal shells is discussed. The new results provided from this work can be used as benchmarks examples for researchers and design engineers.
6) The research work to evaluate the effect of initial pressure for vibration analysis of an inflated toroidal shell is reviewed. And the control analyses of toroidal shells are also reviewed from this work.

ABOUT AUTHOR

Xiaohong Wang is an Associate Professor in the Department of Civil Engineering at the Shantou University. She received her Ph. D. degree from the University of Waterloo in 2008. Her research interests include structural and shell analysis and design.

REFERENCES

ADINA verification manual: ARD2003-9, Watertown (MA, USA), ADINA R and D Inc., 2003.

Armenakas, A.E., Gazis, D.C. and Herrmann, G., Free Vibrations of Circular Cylindrical Shells, Pergamon, Oxford 1969.

Artioli, E. and Viola, E., Free vibration analysis of spherical caps using a GDQ numerical solution, *ASME Journal of Pressure Vessel Technology* 2006,128,370-378.

Artioli, E., Gould, P.L. and Viola, E., A differential quadrature method solution for shear-deformable shells of revolution, *Engineering Structure* 2005,27,1879-1892.

Bacon, M and Bert, C.W., Unsymmetric Free Vibrations of orthotropic Sandwich Shells of Revolution, *AIAA Journal* 1967,5(3), 413-417.

Balderes, T. and Armenakas, A.E., Free vibrations of ring-stiffened toroidal shells. *AIAA Journal* 1973,11,1637-1644.

Bathe, K.J., Finite Element Procedures, Englewood Cliffs, NJ, Prentice-Hall, 1996.

Bert, C.W. and Malik, M., Differential quadrature method in computational mechanics, A review, *Applied Mechanics Review* 1996a, 49(1),6-27.

Bert, C.W. and Malik, M., Free vibration analysis of thin cylindrical shells by the differential quadrature method, *Journal of Pressure Vessel Technology*, 1996b,118,7-12.

Buchanan, G.R. and Liu, Y.J., An analysis of the free vibration of thick-walled isotropic toroidal shells, *International Journal of Mechanical Science* 2005,47,277-292.

Buchanan, G.R. and Rich, B.S., Effect of boundary conditions of free vibration of thick isotropic spherical shells, *Journal of Vibration Control*, 2002,8,389–403.

Budiansky, B., Notes on nonlinear shell theory, *Journal of Applied Mechanics* 1968,40,393-401.

Chen,W.Q. and Ding,H.J., Free vibration of multi-layered spherically isotropic hollow spheres, *International Journal of Mechanical Sciences* 2001,43,667–680.

Chen,W.Q., Effect of radial inhomogeneity on natural frequencies of an anisotropic hollow sphere, *Journal of Sound and Vibration* 1999, 226 (4),787–794.

Chou,F.H. and Achenbach, J.D., Three-dimensional vibrations of orthotropic cylinders, *Journal of Engineering Mechanics* 1972, 813–821.

Clark, R.A., On the theory of thin elastic toroidal shells, *Journal of Mathematics and Physics* 1950, 29(3),146-178.

Ding, H.J. and Chen, W.Q., Natural frequencies of an elastic spherically isotropic hollow sphere submerged in a compressible fluid medium, *Journal of Sound and Vibration* 1996,192 (1),173–198.

Dym, C.L., Introduction to the Theory of Shells, Pergamon, Oxford 1974.

Flügge, W., Stresses in Shells, Springer-Verlag, New York 1967.

Gabbert, U. and Tzou, H.S., Editors, Smart Structures and Structronic Systems, Kluwer Academic Publishers, Dordrecht/Boston/London 2001.

Gautham, B.P. and Ganesan, N., Free vibration analysis of orthotropic thick shells of revolution using discrete layer theory, *Journal of Sound and Vibration* 1994, 171,549–556.

Gavelya, S.P. and Kononenko, N.I., Characteristic oscillations and waves on a toroidal shell, *Soviet Applied Mechanics* 1975, 9(11),31-36.

Griffith, D. T. and Main, J. A., Modal Testing of an inflated thin film polyimide torus structures, *Proceedings of 18th International Modal Analysis Conference* 2000.

Huang, D.W., Redekop, D. and Xu B., Natural frequencies and mode shapes of curved pipes, *Computers and Structures* 1997,63,465-473.

Jha, A.K. and Inman, D.J., Importance of geometric non-linearityand follower pressure load in the dynamic analysis of a gossamer structure, *Journal of Sound and Vibration* 2004, 278, 207–231.

Jha, A.K. and Inman, D.J., Optimal sizes and placements of piezoelectric actuators and sensors on an inflated toroidal shell, *Proceedings of SPIE's 9th Annual Smart Materials and Structures Conference*, March 17-21, 2002b, San Diego, CA.

Jha, A.K. and Inman, D.J., Optimal sizes and placements of piezoelectric actuators and sensors for an inflated torus, *Journal of Intelligent Material Systems and Structures* 2003,14(9),563-576.

Jha, A.K. and Inman, D.J., Piezoelectric actuator and sensor models for an inflated toroidal shell, *Mechanical Systems and Signal Processing* 2002a,16(1), 97-122.

Jha, A.K. and Inman, D.J., Vibration control of a gossamer toroidal structure using smart actuators and sensors, *Proceedings-International Society for Optical Engineering* 2003,5056,475-486.

Jha, A.K., Inman, D.J. and Plaut, R.H., Free vibration analysis of an inflated toroidal shell, *Journal of Vibration and Acoustics* 2002, 124(3), 387-396.

Jiang, W, Hu, X.J. and Redekop, D., Vibration Analysis of a Toroidal Shell Using MAPLE. *Proceeding of 20th CANCAM, Montreal* 2005,154-155.

Jiang, W. and Redekop, D., Polar axisymmetric vibration of a hollow toroid using the differential quadrature method, *Journal of Sound and Vibration* 2002,251(4),761-765.

Jiang, W. and Redekop, D., Static and vibration analysis of orthotropic toroidal shells of variable thickness by differential quadrature, *Thin-Walled Structure* 2003,41,461-479.

Jordan, P.E., Vibration and bucking of pressurized torus shells, *4th Aerospace Science Meeting, Los Angeles, California, AIAA Paper* 1966, No. 66-445.

Kang, J.H. and Leissa, A.W., Free vibration analysis of complete paraboloidal shells of revolution with variable thickness and solid paraboloids from a three- dimensional theory, *Computer and Structure* 2005,83,2594–608.

Kang, J.H. and Leissa, A.W., Three-dimensional vibrations of thick spherical shell segments with variable thickness, *International Journal of Solids and Structures* 2006,43(9),2848-2851.

Kosawada T, Suzuki K, Takahashi S. Free vibrations of thick toroidal shells, *Bull Japan Society of Mechanical Engineering* 1986, 29 (255), 3036-3042.

Kosawada, T., Suzuki. K. and Takahashi, S., Free vibrations of thin toroidal shells, *Bull Japan Society of Mechanical Engineering* 1985,28 (243), 2041-2047.

Kraus, H., Thin Elastic Shell, John Wiley, New York 1967.

Leissa, A. W. and So, J., Three-dimensional vibration of truncated hollow cones, *Journal of Vibration and Control* 1995,1,145–158.

Leissa, A.W. and Kang, J.H., Three-dimensional vibration analysis of thick shells of revolution, *ASCE Journal of Engineering Mechanics* 1999,125,1365-1371.

Leissa, A.W. and Kang, J.H., Three-dimensional vibrations of thick spherical shell segments with variable thickness, *International Journal of Solids and Structures* 2002,37, 4811–4823.

Leissa, A.W., Vibration of Shells, NASA, SP-288, 1973.

Leung, A.Y.T. and Kwok, N.T.C., Dynamic stiffness analysis of toroidal shell, *Thin-Walled Structure* 1995,21,43-64.

Leung, A.Y.T. and Kwok, N.T.C., Free vibration analysis of a toroidal shell, *Thin-Walled Structure* 1994,18, 317-332.

Lewis, J.A., and Inman, D.J., Finite element modeling and active control of an inflated torus using piezoelectric devices, *Journal of Intelligent Material Systems and Structures* 2002,12,819-833.

Lewis, J.A., Finite Element Modeling and Active Control of an Inflated Torus Using Piezoelectric Devices, *MS Thesis, Virginia Polytechnic Institute and State University, Blacksburg, VA* 2000.

Li, S. and Cook, J., An analysis of filament overwound toroidal pressure vessels and optimum design of such structures, *Journal of Pressure Vessel Technology* 2002,124,215-222.

Librescu, L., Elastostatics and kinetics of anisotropic and heterogeneous shell-type structures, Leyden: Noordhoff International Publishing, 1975.

Liepins, A.A., Free vibration of prestressed toroidal membrane, *AIAA Journal* 1965,3(10),1924-1933.

Liew, K.M., Berman, L.A., Ng, T.Y. and Lam, K.Y., Three-dimensional vibration of cylindrical shell panels—solution by continuum and discrete approaches, *Computational Mechanics* 2000, 26,208–221.

Liu, B. and Tzou, H.S., 1998, Distributed Photostrictive Actuation and Opto-piezothermoelasticity Applied to Vibration Control of Plates, *ASME Journal of Vibration and Acoustics* 1998,120,937-946.

Lur'e, A.I., Three-Dimensional Problems of the Theory of Elasticity, Wiley, New York, NY 1964.

Markus, S., The Mechanics of vibrations of cylindrical shells, Elsevier: Amsterdam, 1988, 96.

McGee, O.G. III. And Kim, J.W., Three-dimensional vibrations of cylindrical elastic solids with V-notches and sharp radial cracks, *Journal of Sound and Vibration* 2010,329, 457–484.

McGill DF, Lenzen KH. Polar axisymmetric free oscillations of thick hollowed tori. *SIAM Journal of Applied Mathematics*, 1967,15(3), 679-692.

Ming, R.S., Pan, J. and Norton, M.P., Free vibrations of elastic circular toroidal shells, *Applied Acoustics* 2002,63,513-528.

Noor, A. K. and Peters, J. M., Stress and vibration analyses of anisotropic shells of revolution, *International Journal for Numerical Methods in Engineering* 2005,26 (5),1145-1167.

Novozhilov, V.V., the *Theory of Thin Shells*, Noordhoff, the Netherlands 1959.

Ornyak, I.V., Radchenko, S.A. and Batura, A.S., Calculation of natural and forced vibrations of a piping system, Part 2, Dynamic stiffness of a pipe bend, *Strength of Materials* 2007, 39,144-158.

Park, G., Kim, M. and Inman, D.J., Integration of smart materials into dynamics and control of inflatable space structures, *Journal of Intelligent Material Systems and Structures* 2001, 12,423-433.

Park, G., Sausse, M. and Inman, D.J., Vibration Testing and Finite Element analysis of an inflatable structure, *AIAA Journal* 2003,41(8),1556-1563.

Redekop, D. and Muhammad, T., Analysis of toroidal shells using the differential quadrature method, *International Journal of Structural Stability and Dynamics* 2003, 3(2), 215-226.

Redekop, D. and Xu, B., Dynamic buckling of toroidal shells subject to impulsive local loading, *Proceeding of 9th International Conference Pressure Vessel Technology*, Sydney 2000,1, 685-692.

Redekop, D., Dynamic response of a toroidal shell panel to blast loading, *Computers and Structures* 1994,51, 235-239.

Redekop, D., Dynamic response of curved pipes, *International Journal of Pressure Vessels and Piping* 1997, 70, 167-172.

Redekop, D., Dynamic response of short curved pipes to impulsive loading, *International Journal of Pressure Vessels and Piping* 1995,61,41-47.

Redekop, D., Free Vibration of an orthotropic hollow body of revolution, International *Journal of Structural Stability and Dynamics* 2005,5, 299-312.

Redekop, D., Free Vibration of hollow bodies of revolution, *Journal of Sound and Vibration* 2004a, 273, 415-420.

Redekop, D., Impulsive Local Loading of a Curved Pipe. *Proceedings of the ASME Pressure Vessels and Piping Conference*, New Orleans 1992, 235, 179-183.

Redekop, D., Natural frequencies of short curved pipe, *Transaction of Canadian Society for Mechanical Engineer* 1994, 18(1), 35-45.

Redekop, D., Three-dimensional free vibration analysis of inhomogeneous thick orthotropic shells of revolution using differential quadrature, *Journal of Sound and Vibration* 2006, 291, 1029-1040.

Redekop, D., Vibration analysis of a torus-cylinder assembly, *Journal of Sound and Vibration* 2004b, 277, 919-930.

Redekop, D., Xu, B. and Zhang, Y.M., Vibration of prestressed toroidal panels using the DQM, *Proceeding of ASEM '99 Conference*, Seoul 1999,1, 335-340.

Redekop, D., Xu, B., Vibration analysis of toroidal panels using the differential quadrature method, *Thin-Walled Structures* 1999, 34, 217-231.

Ren, W.M., Liu, W.G., Zhang, W., Reimerdes, H.g. and Oery, H., Survey of works on the theory of toroidal shells and curved tubes, *Acta Mechanica Sinica* (English Series) 1999,15(3), 225-234.

Rossetto, J.N. and Sanders, J.N., Toroidal shell under internal pressure in the transition range, *AIAA Journal* 1965, 3(10),1901-1909.

Ruggiero, E.J. and Inman, D., Gossamer Spacecraft: Recent Trend in Design, Analysis, Experimentation, and Control, *Journal of Spacecraft and Rockets* 2006, 43(1),10-24.

Ruggiero, E.J., Jha, A., Park, G. and Inman, D., A literature Review of Ultra-Light and Inflated Toroidal Satellite Components, *The Shock and Vibration Digest* 2003,53 (3),171-176.

Saigal, S. and Yang, T.Y., Free vibrations of a tire as a toroidal membrane, *Journal of Sound and Vibration* 1986, 107(1), 71-82.

Sanders, J.L. and Liepins, A.A., Toroidal membrane under internal pressure, *AIAA Journal* 1963,1(9) 2105-2110.

Sanders, J.L., *An improved first-approximation theory for thin shells Langley Research Center*, TR R-24, 1959.

Sen, S.K. and Gould, P.L., Free Vibration of Shells of Revolution Using FEM, *Journal of the Engineering Mechanics Division* 1974,100(2), 283-303.

Shu, C., *Differential Quadrature and Its Application in Engineering,* Springer 2000.

Silva, C.N., The effect of transverse shear deformation on the bending of elliptic toroidal shells, *Journal of the Society for Industrial and Applied Mathematics* 1964,12(2),465-476.

Singal, R.K. and Williams, K., A theoretical and experimental study of vibrations of thick circular cylindrical shells and rings, *Journal of Vibration, Stress, and Reliability in Design* 1988, 110, 533–537.

So, J.Y. and Leissa, A.W., Free vibrations of thick hollow circular cylinders from three-dimensional analysis, *Journal of Vibration and Acoustics* 1997,119,89–95.

Soedel, W., On the vibration of shells with Timoshenko-Mindlin type shear deflections and rotary inertia, *Journal of Sound and Vibration* 1982, 83(1),67-79.

Soedel, W., Vibration of Shells and Plates. 3rd. Ed. New York, Marcell Decker 2004.

Tan, D.Y., Free vibration analysis of shells of revolution, *Journal of Sound and Vibration* 1998, 213,15-33.

Timoshenko, S. and Woinowsky-Krieger, S., *Theory of Plates and Shells.* New York, McGraw-Hill 1959.

Tzou, H.S. and Bergman, L.A.(Ed.), Dynamics and Control of Distributed System, *Cambridge University Press*, New York 1998.

Tzou, H.S. and Wang, D.W., Micro-sensing characteristics and modal voltages of linear/non-linear toroidal shells, *Journal of Sound and Vibration* 2002, 254(2), 203-218.

Tzou, H.S., and Wang D.W., Vibration control of toroidal shells with parallel and diagonal piezoelectric actuators, *Journal of Pressure Vessel Technology-Transactions of the ASME* 2003,125(2),171-176.

Tzou, H.S., Bao, Y. and Venkayya, V.B., Study of Segmented Transducers Laminated on Cylindrical Shells, Part 2: Actuator Patches, *Journal of Sound and Vibration* 1996,197(2), 225-249.

Tzou, H.S., Bao, Y. and Zhou, Y., Nonlinear Piezothermoelasticity and Multi-Field Actuations, Part-1: Nonlinear Anisotropic Piezothermoelastic Shell Laminates: Part-2: Control of Nonlinear Buckling and Dynamics, *ASME Journal of Vibration and Acoustics* 1997,119, 374-389.

Tzou, H.S., Chai, W.K. and Wang, D.W., Micro-control actions and location sensitivity of actuator patches laminated on toroidal shells, *Journal of Vibration and Acoustics* 2003,126(2),284-297.

Tzou, H.S., Ding, J.H., and Hagiwara, I., Micro-control Actions of Distributed Actuators Laminated on Precision Paraboloidal Structronic Shells, *JSME International Journal, a Special Issue on Dynamics and Design of Continuous Systems, Series C*, 2002,45(1), 8-15.

Tzou, H.S., Zhong, J.P. and Hollkamp, J.J., Spatially Distributed Orthogonal Piezoelectric Shell Actuators: Theory and Applications, *Journal of Sound and Vibration* 1994, 177(3), 363-378.

Wang, X.H. and Redekop, D., Natural frequencies and mode shapes of an orthotropic thin shell of revolution, *Thin-Walled Structure* 2005, 43, 735-50.

Wang, X.H., Free vibration and stability of complete toroidal shells, *MASc Thesis*, University of Ottawa, 2004.

Wang, X.H., Xu, B. and Redekop, D., FEM free vibration and buckling analysis of stiffened toroidal shells, *Thin-Walled Structure* 2006, 44, 2-9.

Wang, X.H., Xu, B. and Redekop, D., Theoretical natural frequencies and mode shapes for thin and thick curved pipes and toroidal shells, *Journal of Sound and Vibration* 2006, 292, 424-434.

Xu, B. and Redekop, D., Natural frequencies of an orthotropic thin toroidal shell of elliptical cross-section, *Journal of Sound and Vibration* 2006, 293, 440-448.

Yamada, G., Kobayashi, Y., Ohta, Y. and Yokota,S., Free vibration of a toroidal shell with elliptical cross-section, *Journal of Sound and Vibration* 1989, 135,411-425.

Zhan, H.J. and Redekop, D., Projectile Impact on a Toroidal Shell, *5th International Conference on Thin-Walled Structures*, Brisbane, Australia 2008, 843-850.

Zhan, H.J. and Redekop, D., Vibration, buckling and collapse of ovaloid toroidal tanks, Thin-Walled Structure 2008, 46, 380-389.

Zhou, D., Cheung, Y.K., Lo, S.H. and Au, F.T.K., 3D vibration analysis of solid and hollow circular cylinders via Chebyshev-Ritz method, *Computer Methods in Applied Mechanics and Engineering* 2003,192,1575–1589.

Zhou, Y.H. and Tzou, H.S., Control of Nonlinear Piezoelectric Circular Shallow Spherical Shells, *Journal of Solids and Structures* 2000,37,1663-1677.

INDEX

A

actuators, 174, 177, 180
aerospace, 147
Africa, 21
ANSYS program, 133, 134, 135
assessment, 147
attachment, 148

B

base, 9
Beijing, 23, 75
benchmarks, 169, 175
bending, 3, 4, 6, 13, 22, 69, 78, 79, 80, 81, 82, 117, 118, 119, 122, 126, 128, 129, 130, 148, 151, 155, 159, 160, 180
biomechanics, 147
bonds, 4
breakdown, viii, 117

C

carbon, 4, 21, 22, 24
carbon atoms, 4
carbon nanotubes, 4, 22
case study(s), 135, 147, 163, 169
challenges, 173
China, 23, 25, 75, 76, 151, 152
classes, 147
classification, 105
cobalt, 77, 106, 107, 108, 109, 110
commercial, 136, 139, 147, 169, 171, 172
communication, 173
compatibility, 2, 28, 29, 30
compensation, 11
complement, 147
complexity, 8
compliance, 106
compression, 4
computer, 80, 112, 158, 169, 174
computer software, 158, 174
condensation, 153
configuration, 174
construction, 147
containers, vii, 1, 152
cooling, 173
covering, 157
cracks, 178
critical value, 17
crown, 154
crystal structure, 107
crystals, 113
curved pipes, vii, viii, 1, 23, 117, 118, 128, 148, 149, 152, 171, 172, 176, 179, 181
cylindrical coordinates, 78, 99, 105, 110, 111
Czech Republic, 22

D

deformation, 2, 3, 15, 21, 22, 53, 54, 55, 66, 67, 81, 111, 115, 117, 118, 119, 120, 159, 160, 175
depth, 92, 111
derivatives, 70, 73, 92, 157
diaphragm, 152, 169, 170
differential equations, 111, 122, 156, 157, 162
dilation, 120
displacement, viii, 3, 5, 8, 11, 13, 15, 18, 19, 20, 41, 78, 81, 83, 84, 86, 92, 93, 94, 99, 100, 116, 117, 118, 120, 121, 122, 123, 124, 125, 126, 127, 136, 138, 139, 141, 142, 143, 145, 146, 153, 156, 157, 158, 161, 162, 166, 167, 174
distortions, 118
distributed load, 26, 27
distribution, 14, 98, 146

E

elasticity modulus, 132
elbows, vii, 1, 148
electric field, 75
elliptical coordinates, 102
elliptical toroidal shells, 102, 103
e-mail, 25, 115, 151
energy, 82, 83, 95
engineering, vii, 1, 9, 80, 105, 106, 117, 152, 153, 159, 164, 169
equality, 36
equilibrium, 2, 27, 30, 161
evolution, 145
excitation, 174
extraction, 158

F

fabrication, 118
FEM, 23, 153, 157, 158, 163, 164, 165, 168, 169, 170, 171, 172, 174, 175, 179, 180
fiber, 112, 152
filament, 178
finite element method, viii, 79, 80, 111, 112, 175
flexibility, viii, 117
fluid, 27, 39, 176
force, 11, 18, 27, 28, 34, 38, 39, 50, 55, 67, 117, 120, 130, 132, 137, 138, 139, 140, 141, 142, 143, 160
formula, 50, 52, 59, 69, 71, 73, 74
Fourier expansion displacement field, viii, 117
Fourier harmonic index, vii, 25, 26
freedom, 118, 121, 122, 123, 126, 169
friction, 14
full toroidal shells, vii, 1
fusion, vii, 1, 169

G

geometrical parameters, 54
geometry, 84, 113, 118, 128, 154, 157, 158, 163, 169, 170, 171, 172
Germany, 152

H

hybrid, 148

I

ideal, 4
identity, 53
impulsive, 179
inertia, viii, 79, 80, 84, 87, 90, 96, 101, 111, 112, 113, 154, 161, 180
inhomogeneity, 176
integration, 10, 11, 12, 15, 16, 17, 53, 61, 70, 89, 121, 148, 174
isotropic media, 113
Israel, 21, 22, 23
iteration, 170

J

Japan, 177
joints, 21
Jordan, 173, 177

K

kinetics, 178

L

laws, 4
lead, 82
light, vii, 1, 23
long piping elbows, vii, 1

M

machinery, 164
magnetic moment, 4
mass, 78, 79, 83, 84, 89, 94, 100, 107, 110, 155, 158, 161, 167, 170
materials, 4, 14, 75, 80, 93, 105, 106, 107, 110, 112, 147, 160, 161, 165, 166, 174
mathematics, 147
matrix, 4, 14, 77, 78, 79, 80, 83, 84, 87, 88, 89, 93, 94, 95, 99, 100, 101, 105, 106, 107, 108, 109, 110, 115, 116, 117, 120, 121, 127, 128, 152, 157, 158
matter, viii, 116
mechanical properties, 3, 4, 21
memory, 75
meridian, vii, 1, 25, 66, 79, 81, 154, 159, 162, 163, 173, 175
methodology, 97, 135

modelling, 4, 117, 147
models, 4, 147, 174, 177
modulus, 4, 30, 78, 89, 95, 107, 115, 121, 158, 166, 170, 171, 172
molecular dynamics, 22
molecules, 5
Mushtari-Vlasov-Donnell (MVD), 1

N

nanocomposites, 4
nanostructures, 4
nanotube, 4, 14, 22, 23
neglect, 38
NEMS, 4
Netherlands, 178
nodes, 78, 89, 118, 121
numerical analysis, 152

O

operations, 146
orbit, 173
ordinary differential equations, 9, 16, 48, 117
oxygen, vii, 1

P

parallel, 28, 78, 79, 81, 174, 180
partial differential equations, 83
permit, 32
personal computers, 146
Philadelphia, 112
physics, 105
piezoelectricity, 174
Poisson ratio, 155
polyimide, 176
polymer, 4
polynomial functions, 118
Portugal, 115, 147
preparation, 112
project, viii, 112, 116, 117, 128
PVP, 22, 23

R

radius, 4, 6, 7, 18, 25, 31, 77, 78, 81, 85, 86, 89, 91, 92, 95, 96, 97, 98, 99, 103, 107, 110, 116, 118, 119, 135, 136, 137, 153, 154, 158, 159, 163, 165, 166, 168, 170, 171, 172, 173
reference system, 124
reinforcement, vii, 1
researchers, vii, viii, 79, 102, 117, 169, 175
response, 24, 173, 179
rings, 174, 180
risk, viii, 117
rotations, 12, 28, 29, 30, 78, 79, 81, 82, 86, 122, 126, 160, 161, 162
routines, 158
Royal Society, 21

S

safety, viii, 116, 173
scaling, 164
scope, 2
senses, 130
sensing, 174, 180
sensitivity, 180
sensors, 174, 177
shape, 2, 4, 77, 78, 82, 83, 84, 88, 94, 95, 100, 101, 116, 117, 121, 123, 125, 127, 136, 165
shear, viii, 52, 78, 79, 80, 81, 82, 84, 87, 89, 90, 95, 96, 101, 105, 107, 111, 112, 113, 116, 120, 121, 132, 153, 154, 159, 160, 161, 162, 163, 166, 169, 175, 176, 180
shear deformation, viii, 79, 80, 81, 84, 87, 90, 96, 101, 111, 112, 121, 153, 159, 160, 162, 163, 169, 175, 180
showing, 86, 165
simulation, 4, 5, 22, 117, 147, 174
smart materials, 174, 178
software, 147, 174
South Africa, 1, 21, 23
SP, 177
SSA, 21
stability, 2, 27, 39, 75, 118, 180
state, viii, 16, 80, 117, 118, 151
steel, 4, 132, 135, 137, 147, 148
storage, 159
stress, viii, 4, 24, 30, 39, 40, 50, 52, 81, 106, 107, 108, 116, 117, 118, 120, 121, 128, 130, 131, 132, 146, 148, 151, 153, 155, 160, 166, 173
structural engineering, vii, 1
structure, viii, 4, 106, 116, 117, 118, 121, 132, 133, 135, 137, 140, 142, 174, 177, 178
submarines, vii, 1
substitution, 31, 46, 47, 54, 63
Sun, v, 1, 3, 4, 20, 21, 23, 112
supervisor, 75
suppression, 174
symmetry, 2, 128, 169

T

tanks, vii, 1, 152, 159, 181
techniques, 105, 117, 147, 148
technology, 152, 159
temperature, 116, 120, 126, 127, 135, 136, 137, 138, 139, 141, 142, 143, 144, 145
tensile strength, 4
tension, 4
testing, 174
textbooks, 80, 105
theoretical approaches, 175
thermal deformation, 120
thermal expansion, 11, 20, 116, 120, 127, 135, 136, 137
torsion, 4
torus, 4, 77, 80, 85, 91, 92, 97, 101, 102, 113, 173, 174, 176, 177, 178, 179
transformation, 1, 2, 3, 8, 22, 41, 56, 67, 110, 111
translation, 21, 22, 23, 75
transverse section, 118, 128, 133, 146
trial, 157

U

UK, 149
underwater toroidal pressure hulls, vii, 1
uniform, 21, 54, 55, 128, 137, 140, 141, 143, 144, 151
USA, 77, 113, 175

V

validation, 158, 164, 168, 172
variables, 30, 34, 36, 39, 82, 83, 95, 107, 160, 161, 162, 167
variational method, viii, 117
vector, 10, 16, 78, 79, 85, 116, 120, 126, 127, 130, 153, 154, 159, 170
versatility, 99, 102
vessels, vii, 1, 23, 117, 118, 148, 152, 159, 164, 169, 178
vibration, vii, viii, 3, 15, 20, 21, 22, 23, 24, 80, 81, 82, 83, 87, 88, 89, 90, 91, 94, 95, 98, 100, 102, 110, 111, 112, 113, 151, 152, 153, 154, 155, 157, 158, 159, 160, 165, 169, 170, 172, 173, 174, 175, 176, 177, 178, 179, 180, 181

W

waste, 152
water, 159
wave number, vii, 26, 78, 88, 94, 100
wave propagation, 113
workers, 102, 152

Y

yield, 35

Z

zinc, 106, 107, 110